If found, Please return t
ANTHONY FORMOSA
Centrica Energy
Millstream West
Maidenhead Road
Windsor
SL4

Errata   Underground Gas Storage Facilities

Dear Reader,

Unfortunately, errors do occur in books. We apologize for the inconvenience, and if you have any questions whatsoever, please write directly to me.

William J. Lowe
Editor-in-Chief, Book Division

On page 12, Table 2-1 is actually Figure 2-1. Table 2-1 should read as follows:

Table 2-1
Example of Pro Forma Adjustment in
Number of Residential Customers

Year	Actual No. of Customers	Pro Forma Adjustment			Pro Forma Customers
		No. 1	No. 2	Total	
1975	149,457	5,036	(5,926)	(890)	148,567
1976	149,509	5,154	(2,518)	2,636	152,145
1977	151,227	5,272	0	5,272	156,499
1978	155,702	5,390	0	5,390	161,092
1979	159,056	5,508	0	5,508	164,564
1980	161,476	5,626	0	5,626	167,102
1981	164,620	5,744	0	5,744	170,364
1982	167,988	5,862	0	5,862	173,850
1983	175,453	997	0	997	176,450
1984	178,510	0	0	0	178,510
1985	180,856	0	0	0	180,856
1986	183,202	0	0	0	183,202
1987	185,042	0	0	0	185,042

On page 32, paragraph 3, "Bernoulli" should be "Bordon."

On page 85, line 8 of the second paragraph, "transmission" should be "transmission line."

On page 113, the right-hand side of Equations 9-6 and 9-7 should have a minus sign.

On page 153, the left-hand side of Equation 12-3 should have a "k" in the denominator.

UNDERGROUND GAS STORAGE FACILITIES

UNDERGROUND GAS STORAGE FACILITIES

Design and Implementation

Orin Flanigan

Gulf Publishing Company
Houston, London, Paris, Zurich, Tokyo

Underground Gas Storage Facilities
Design and Implementation

Copyright © 1995 by Gulf Publishing Company, Houston, Texas. All rights reserved.

Gulf Publishing Company
Book Division
P.O. Box 2608 ☐ Houston, Texas 77252-2608

10 9 8 7 6 5 4 3 2 1

Library of Congress Cataloging-in-Publication Data

Flanigan, Orin.
    Underground gas storage facilities : Design and implementation / Orin Flanigan.
       p. cm.
    Includes bibliographical references and index.
    ISBN 0-88415-204-9 (acid-free)
    1. Natural gas—Underground storage.   I. Title.
TP756.5.F57   1995
665.7'42—dc20   94-49706
                                                               CIP

Transferred to digital printing 2006

This book is dedicated first and foremost to my wife, Billie, who makes life interesting. It is also dedicated to that group of old guard Arkla employees with whom I have associated for many years and who have made being in the pipeline business an experience to remember.

Contents

Preface .... **x**

CHAPTER 1
Weather Patterns **1**
    Fuel Storage 1. Historical Weather Data 3.

CHAPTER 2
Load Forecasting **9**
    Types of Customers 10. Annual Average Number of Customers 11. Monthly Number of Customers 13. Usage Per Customer 16. Non-Consumer Gas Requirements 19. Total Requirements 19. Daily Weather Forecasting 21. References 24.

CHAPTER 3
Load Curves **25**
    Requirements Load Curves 25. Requirements and Supply Load Curves 26. Supply and Storage Load Curves 28.

CHAPTER 4
Gas Laws **32**
    Pressure Measurement 32. Temperature Measurement 33. The Effect of Pressure on Gas Volume 33. The Effect of Temperature on Gas Volume 34. Ideal Gas Law 36. Compressibility Factors 37. Approximations 38.

CHAPTER 5
The Components of a Gas Storage Facility **40**
    Underground Reservoirs 40. Wells 44. Gathering System 48. Compressors 49. Central Point Metering 51. Central Point Separators 51. Central Point Dehydrators 52. Transmission Line 53.

CHAPTER 6
Characteristics of Underground Storage 54
Storage Characteristics 54. Types of Storage 59. Tailoring of Storage Facilities 61. Sizing of Surface Facilities 64.

CHAPTER 7
Optimization of Underground Storage Facilities 68
Storage Wells 69. Working Gas Volume 72. Injection Horsepower 76. Types of Storage 77. Optimization 80. Summary 89.

CHAPTER 8
Monitoring and Control of Inventory 90
Types of Leakage 92. Pressure-Volume History 92. References 108.

CHAPTER 9
Pressure Measurements on Reservoirs 109
Background 109. Theoretical Development 110. Example 114. References 117.

CHAPTER 10
Metering 118
Types of Meters 118. Sources of Meter Error 119. Current Research 121. Pulsating Flow 121. References 131.

CHAPTER 11
Dehydration 132
Hydrates 132. Water Content of Natural Gas 133. Types of Dehydrators 133. Glycol Dehydrators 134. Dew Point Measurement 140. Water Content Charts 142. Tabulating Water Content Data 144. References 146.

CHAPTER 12
Compressors 147
Injection versus Withdrawal Requirements 147. Lubrication Options 148. Two-cycle versus Four-cycle Engines 149. Control Technology 149. Inlet Separator 149. Fin Fan Coolers 150. Preventative Maintenance 151. Compressor Models 151. References 156.

CHAPTER 13
Estimating Deliverability of Producing Wells 157
Historical Deliverability Forecasting 157. Proposed Approximate Method 162. References 166.

CHAPTER 14
Automation 167

APPENDIX A
Evaluation of Constants for Approximate Compressibility Factor Equation 170

APPENDIX B
Development of Compressor Horsepower Equation 174

Index 176

Preface

Most books on the underground storage of natural gas are written from the viewpoint of the petroleum engineer or the reservoir engineer. This is natural because these two disciplines are heavily involved in evaluating the reservoir for the prospective project. This book assumes that the petroleum engineer and the reservoir engineer have already done their job. One or more reservoirs have been selected, their physical properties such as porosity and permeability have been determined, and their aerial extent, depth, and thickness have been measured. The security of the reservoirs has been evaluated and a judgment made that they are sufficiently secure to proceed.

At this point the pipeline engineer and the planner enter the scene. These individuals are fully familiar with the transmission facilities and their operation as well as the characteristics of the various gas supplies and the customer loads. They are also familiar with the economic and operating philosophies peculiar to their particular company. The purpose of this book is to provide these individuals with the information and technical tools necessary to interact with the petroleum engineer and the reservoir engineer, to evaluate each reservoir, and to integrate this information into the design of a storage facility.

UNDERGROUND GAS STORAGE FACILITIES

CHAPTER 1

Weather Patterns

Energy is a vital part of practically all industrial concerns. Without it, most companies would cease to function. This energy can originate from many basic sources including:

        Gas   Water power
        Oil   Wind power
        Coal   Solar power
        Nuclear power   Geothermal power

Electricity is missing from this list because electricity is a secondary source of power. All of the above sources of fuel can be converted into electricity.

Nuclear power is so regulated and so financially demanding that it is practical only for the production of power in very large amounts. It is not viable for the ordinary medium-sized industrial organization. Water power is widely used for the production of electricity, but it is no longer used to any extent as a source of industrial power by itself. Wind power and solar power have had some small development in special situations, but they are not practical for the average fuel user. Geothermal power is severely limited by its geographic occurrence. This leaves coal, oil, and gas as the primary sources of fuel used by industry.

Fuel Storage

Any fuel requires storage to some degree. Coal is delivered from the mine to the point of usage by rail or truck. Since these deliveries are not continuous, coal must be stored to accommodate usage between deliveries. This can be done in bins for moderate amounts or in open stock piles for larger amounts. The coal piles are stable, are not unduly affected by the weather, and the storage effort is reasonably economical. Fuel oil (and other liquid fuels) is similar to coal in some of these aspects. Much of the oil is delivered by truck or rail. This requires that storage be available to accommodate usage between deliveries. Fuel oil can be delivered to larger industrial users by pipeline. Even in this case, however, the pipeline is not dedicated to delivering oil alone. Usually the pipeline transports batches of different products. These hydrocarbon products may vary from gasoline on one end of the spectrum to heavy fuel oil on the other. Because of this, the shipments of oil by pipeline are periodic, and storage must be provided for the periods of usage between shipments. This storage may take the form of atmospheric pressure tanks.

Natural gas is the only one of these commonly used fuels that does not require storage by the user. Natural gas is delivered to the customer by pipelines. These pipelines are dedicated to the customers they serve, and they transport only the one product. The user always has a supply of natural gas available for use by simply turning on a valve. Unfortunately, this is not the end of the storage story.

The transmission pipelines that carry the natural gas to the customer have some limit on their capacity. The pipelines also have a cost associated with them. As long as the loads that the pipeline serves are constant throughout the year, the pipelines can be economically designed to handle the service. When a large portion of the customer loads are temperature sensitive, however, a problem arises. Many residential, commercial, and industrial customers use gas for environmental heating. This is heating to counteract the low ambient temperatures that occur in winter. The colder the atmospheric temperature becomes, the greater is the amount of gas required to heat the customer structures. The coldest days of the winter only occur a few days each year. If a large portion of a pipeline's load is heat sensitive, economically designing facilities to supply this load on a year-round basis can be difficult. If the limitation on service during the very cold periods is due to pipeline capacity limitations, gas storage located near the consumption area is necessary. If the limitation on service is due to a supply limitation, the gas storage location is not as critical.

Storage techniques for the three primary fuels differ due to the differing nature of the three fuels. A standard cubic foot of each of the three fuels contains the following heat content in Btus:

Coal	1,100,000
Oil	1,100,000
Natural gas	1,000

Natural gas is the most difficult to store; because it is a gas, it must be contained in a container that is leak-proof under pressure. Neither coal nor oil has this restriction. Because of its lower heating value per unit volume at atmospheric pressure, gas requires either much larger atmospheric pressure storage vessels or pressurized storage vessels.

Fuel gas has been stored in atmospheric pressure storage vessels. These vessels, called gas holders, were used in the early part of this century to store manufactured gas in large cities. The holders had telescoping sections that rose vertically as gas was introduced into the vessels. The pressure in the holders was about 10 to 15 inches of water column. The largest of these gas holders had a capacity of about 1,000,000 standard cubic feet and were roughly 100 feet in diameter and about 100 feet high when full. These storage devices worked well with the relatively low volumes of relatively high-priced gas that were consumed in the area. When larger volumes of natural gas were introduced into the cities and the prices declined, the older gas holders were no longer adequate or economical to continue to use.

Pressure vessels have also been used for gas storage. Because of the economics and logistics, these storage efforts have been limited to relatively small volumes and applied to special cases where the higher cost could be justified. Mixtures of propane and air have been used to supplement natural gas supplies on very cold days. This mixture is considerably more expensive than natural gas, so that its application has been restricted to special situations.

The underground storage of natural gas has evolved as the preferred means of supplementing natural gas supplies for temperature-sensitive loads on very cold days. In most cases a former producing field is used and transformed into a storage facility. In some cases storage facilities have been developed in aquifers that did not previously hold gas. In either case the reservoir and the supporting equipment must be tailored to the needs of the pipeline company.

Historical Weather Data

For heat-sensitive gas usage, accurate load forecasting requires accurate weather information. At the present time, accurate long-range weather forecasting is not feasible. It is possible, however, to determine what temperature pattern should be expected for a normal winter period. This normal pattern can serve as a basis for a load forecasting procedure.

The U.S. Weather Bureau keeps extensive historical temperature records at each of their many weather stations. This information is processed to yield a "normal" temperature for each day of the year for each station. Unfortunately, this weather bureau normal is not adequate for gas load forecasting.

4 *Underground Gas Storage Facilities*

The procedure used by the weather bureau is to take the high temperature and low temperature for each day and average these to yield a daily average temperature for that location. These daily average temperatures are then averaged for each day of the year for some long period, typically 30 years, to yield a normal temperature for that date.

Table 1-1 illustrates this process, although it is an oversimplification of the Weather Bureau averaging process. The Weather Bureau effectively plots the data for each month, smoothes the data, and uses the smoothed data to obtain the averages. Table 1-1 does, however, illustrate the problem that this processed data presents for the load forecaster.

Table 1-1 contains the February temperature history for five consecutive years for a given location. Both the temperature history and the location are hypothetical, and the five years represent no particular period. The average monthly temperatures for each of the five years are similar, and the number of degree days for each of the five February months varies less than 10 percent. The five monthly periods do not represent extremes, but they are quite similar. Each of the five months contains a low temperature of 20°F and a high temperature of 50°F.

An examination of the five-year temperature average in the right-hand column, however, shows that the lowest temperature in this column is 28.2°F. Similarly, the highest temperature is 41.6°F. This is in contrast to the low temperature of 20°F that occurred in each of the five individual years. This is because the 20°F temperature occurred on a different date in each of the five years. The same is true of the 50°F temperature in each year. Thus the average temperature data in the right-hand column of Table 1-1 would be fine for estimating the monthly gas consumption for a heat-sensitive load, but it would be quite inaccurate in showing the extremes of load that could occur on given days.

In order to alleviate this problem, a different processing procedure must be used for the raw data. This can be accomplished by averaging the data for the coldest day in each month, the second coldest day in each month, the third coldest day in each month, and so forth. Table 1-2 illustrates the intermediate step in this process. The temperature data for the month for year 1 is sorted by ascending temperature. The same is done for years 2 through 5. The left-hand column in Table 1-2 no longer represents the day of the month, it now is the number of days that the temperature is as cold

(text continued on page 8)

Weather Patterns 5

Table 1-1
Hypothetical February Temperature History—Raw Data

Day of Month	Daily Average Temperature					5-Year Average Temp.
	Year 1	Year 2	Year 3	Year 4	Year 5	
1	40	25	33	27	29	30.8
2	39	24	34	20	27	28.8
3	33	20	37	26	32	29.6
4	29	25	33	41	20	29.6
5	20	37	29	43	28	31.4
6	29	44	25	41	37	35.2
7	25	43	20	33	35	31.2
8	22	32	39	32	29	30.8
9	24	31	34	29	24	28.4
10	26	35	31	20	29	28.2
11	35	39	25	22	30	30.2
12	42	43	29	34	31	35.8
13	50	47	32	39	29	39.4
14	38	42	25	33	28	33.2
15	35	50	39	29	35	37.6
16	25	39	43	27	37	34.2
17	29	33	50	32	43	37.4
18	35	29	43	34	48	37.8
19	38	28	29	37	50	36.4
20	29	31	28	43	45	35.2
21	43	35	23	48	39	37.6
22	39	39	32	50	37	39.4
23	34	41	34	45	33	37.4
24	41	45	39	44	39	41.6
25	39	33	43	37	37	37.8
26	35	38	34	29	35	34.2
27	37	36	33	27	34	33.4
28	36	35	29	31	35	33.2
Average	33.8	35.7	33.0	34.0	34.1	34.1
Degree Days	873	821	895	867	865	864

Table 1-2
Hypothetical February
Temperature History—Sorted Data

No. of Days	Sorted Daily Average Temperature					5-Year Average Temp.
	Year 1	Year 2	Year 3	Year 4	Year 5	
1	20	20	20	20	20	20.0
2	22	24	23	20	24	22.6
3	24	25	25	22	27	24.6
4	25	25	25	26	28	25.8
5	25	28	25	27	28	26.6
6	26	29	28	27	29	27.8
7	29	31	29	27	29	29.0
8	29	31	29	29	29	29.4
9	29	32	29	29	29	29.6
10	29	33	29	29	30	30.0
11	33	33	31	31	31	31.8
12	34	35	32	32	32	33.0
13	35	35	32	32	33	33.4
14	35	35	33	33	34	34.0
15	35	36	33	33	35	34.4
16	35	37	33	34	35	34.8
17	36	38	34	34	35	35.4
18	37	39	34	37	35	36.4
19	38	39	34	37	37	37.0
20	38	39	34	39	37	37.4
21	39	41	37	41	37	39.0
22	39	42	39	41	37	39.6
23	39	43	39	43	39	40.6
24	40	43	39	43	39	40.8
25	41	44	43	44	43	43.0
26	42	45	43	45	45	44.0
27	43	47	43	48	48	45.8
28	50	50	50	50	50	50.0
Average	33.8	35.7	33.0	34.0	34.1	34.1
Degree Days	873	821	895	867	865	864

Table 1-3
Hypothetical February Temperature History—Fabricated Normal Temperature Pattern

Day of Month	Normal Temp.
1	29.4
2	24.6
3	26.6
4	25.8
5	30.0
6	34.8
7	29.6
8	29.0
9	22.6
10	20.0
11	27.8
12	36.4
13	45.8
14	33.0
15	39.6
16	34.0
17	37.4
18	40.8
19	37.0
20	35.4
21	40.6
22	44.0
23	39.0
24	50.0
25	43.0
26	34.4
27	33.4
28	31.8
Average	34.1
Degree Days	864

(text continued from page 4)

or colder than the indicated temperature. The right-hand column is the average of the sorted temperature data.

The average of the sorted temperatures now shows a considerably wider range of temperature variation than the average of the raw data shown in Table 1-1. In addition, the extreme high and low of 20°F and 50°F of each individual month is preserved. This now represents a normal monthly pattern that more closely resembles what actually happens in the five data periods. The 5-year averages in Table 1-2 do not, however, indicate where during the month the individual temperatures occurred.

The average temperatures in Table 1-1 are helpful in resolving this problem. Although the range of values in the averages of Table 1-1 are not satisfactory, the individual temperature values do show the dates on which the coldest and warmest temperatures occurred. This information can be used to assign the coldest temperature in Table 1-2 to the coldest day in Table 1-1, the second coldest day in Table 1-2 to the second coldest day in Table 1-1, the third coldest day in Table 1-2 to the third coldest day in Table 1-1, and so on. Table 1-3 shows the result of this series of assignments.

The normal temperature pattern shown in Table 1-3 has the same average monthly temperature as the 5-year average of the raw data shown in Table 1-1. It also has the same number of degree days. Thus the character and validity of the monthly data have not been compromised. Only the range of the data has been changed to better reflect reality. This same procedure can be followed for each month of the year.

In order to use this procedure with the Weather Bureau data, it is necessary to obtain the individual daily average temperatures for a given location for the period desired. This data would involve about 10,550 data points for each location to cover a 30-year period (365 × 30). In the past this information was available in printed form, but it should be available on computer tape by now. The above procedure may be used to process the data to give a realistic temperature pattern for an entire year. These temperatures may then be used by the load forecaster to forecast the gas usage for each individual day of the year on a normal basis.

In addition to the normal gas usage, the load forecaster often wants to be able to design facilities to meet a 10-year, 20-year, or 30-year extreme of temperature. The temperature information for this exercise is also available from the Weather Bureau. The Weather Bureau collects high-temperature extremes and low-temperature extremes for each day of the year at each weather station for a given period, typically 30 years. This data may be screened to give a statistical 10-year, 20-year, or 30-year extreme of high and low temperature.

CHAPTER 2

Load Forecasting

In order to effectively and efficiently design and utilize underground storage it is necessary to identify the storage needs. Load forecasting is the basis for this. Gas distribution companies have been using load forecasting methods for many years. Pipeline companies have, to a lesser extent, employed some methods of estimating their gas requirements. Because the pipeline companies rely so heavily on the distribution company loads, it would seem appropriate to examine the methods that are used by distribution companies to estimate their customer loads. In addition, many gas companies are integrated. These companies include in their operations the production, processing, transmission, and distribution of gas. In these cases it would seem logical that the load forecasting for the entire company be done by a central group.

The basic philosophy of load forecasting may be stated by the equation:

Customer Usage = N × U × DD × D   (2-1)

N = number of customers
U = usage per customer in MScf per degree day per day
DD = average number of degree days per day for the period
D = number of days in the time period

This fairly simple equation will be elaborated on further in the following pages.
There is wide diversity in gas usage among various types of customers. Also different geographic regions use different quantities. Rural customers operate differently than urban dwellers. In spite of these differences, however, customers can be grouped for load forecasting. A geographic area as large as a state may be used for a single type of customer as long as a single temperature sensing location will adequately describe the temperature conditions for the entire area. It is essential, however, that the various types of customers be separated and each type grouped together.

Types of Customers

The following represent the major types of customers:

      Residential   Small industrial
      Small commercial   Medium industrial
      Large commercial   Large industrial

Each of these customer types has its own characteristics. Any individual gas company may have some variation of these classifications.

Residential. These customers usually have one or more of the following gas loads:

      Space heating   Water heating
      Cooking   Clothes drying

The space-heating load is highly temperature sensitive. The water-heating load is temperature sensitive to a limited degree, as will be discussed later. The cooking and clothes-drying loads are usually regarded as insensitive to atmospheric temperature. Cooking and clothes drying remain relatively constant week to week, although they may vary somewhat with the seasons. For example, the amount of cooking may be slightly less during the summer when families are cooking out more on charcoal grills, are on vacation, or cook less demanding meals.

Space heating can and does vary with day of the week. In many families there may be no one at home during normal working hours from Monday to Friday. Space-heating usage would be less on these days. On Saturday and Sunday the increased fuel consumption would reflect more people being home. Space heating also varies widely with the seasons.

Small Commercial. Small commercial establishments usually have a space-heating requirement and may have some additional fuel needs. There is usually no significant process load.

Large Commercial. The distinction between large and small commercial customers is usually in the quantity of their gas usage and thereby their rate schedules. They can have significant process loads.

Small Industrial. The criteria for classifying customers as small commercial or small industrial is usually set by each gas company. Their gas usage characteristics are similar.

Medium Industrial. Medium industrial customers usually have a temperature-sensitive load and a significant process load.

Large Industrial. Large industrial customers usually have a very large process load that overshadows any temperature-sensitive load. They also usually have a different rate schedule than the medium industrial customers.

Annual Average Number of Customers

Once the geographic area has been selected for grouping the various types of customers, it is necessary to forecast the annual average number of customers by each type. Company records showing this history should be available. Use the 12 months ending in December for each classification of customer.

The determination of historic customer growth rates is not a simple endeavor. Any "one-time" transaction involving a significant number of customers can distort and hide the true growth rate for several years. Examples of some of these transactions are:

1. Acquisition of a group of customers.
2. Disposition or sale of a group of customers.
3. Transfer of customers between company divisions.
4. Transfer of customers between customer classes (for example, residential to commercial).

Since the historical trends form a basis for forecasting future growth, it is essential that a proper determination be made of the growth rates in the past. In order to accomplish this, the actual information on the average annual number of customers must be accumulated for the past several years. The 12 months ending in December average number of customers is the best data to use. The year-end number of customers should not be used for reasons that will be explained later. In addition, all one-time transactions must be identified and a means developed for adjusting the actual data to eliminate the effect of these transactions.

Table 2-1 shows an example of how these adjustments can be made. The actual number of residential customers is shown for each year. This particular company history had two customer transactions that affected the customer count:

12 *Underground Gas Storage Facilities*

TABLE 2-1
Example of Pro Forma Adjustment in Number of Residential Customers

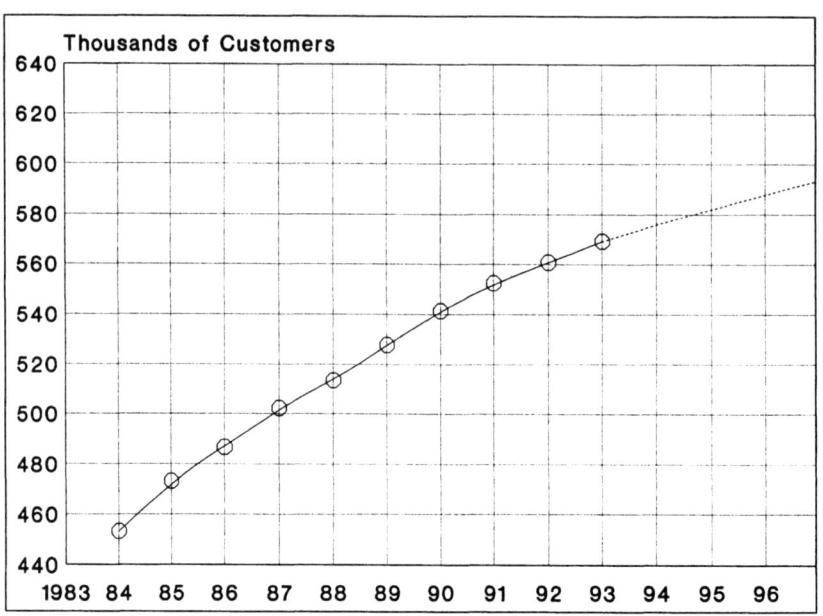

1. On March 1, 1983, a total of 5,980 residential customers were acquired by the purchase of a city distribution system. This small system was determined to have an annual growth rate of 118 customers per year.
2. On May 31, 1976, a total of 6,043 customers in a small isolated distribution system were sold to another gas company. This system was determined to have an annual growth rate of 117 customers per year.

Table 2-1 uses this information to adjust the previous customer history.

Transaction 1 took place after 2 months of the year 1983 had elapsed. Thus the adjustment to the year 1983 was the customer transfer (5,980) times two-twelfths, or 997. The adjustments for the prior years are the 5,980 customers transferred minus the customer growth of 118 customers per year. Similarly, transaction 2 took place after 5 months of 1976 had elapsed. The adjustment for 1976 was the customer transfer of −6,043 times five-twelfths, or −2,518. The adjustment for prior years is the −6,043 plus the 117 customer growth rate per year. The result is shown in the right-hand column.

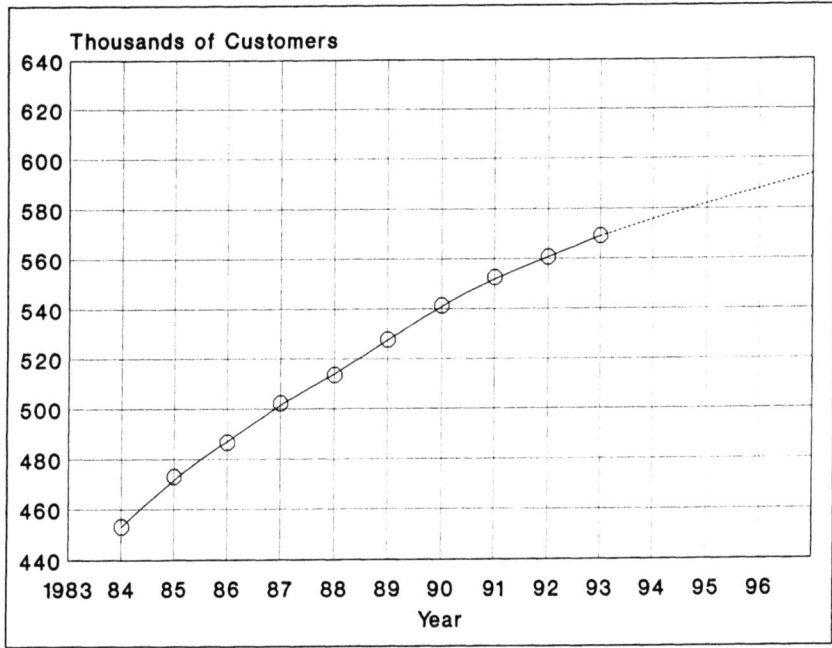

Figure 2-1. Average annual number of residential customers.

Figure 2-1 shows graphically the history of the residential customers for a typical gas company geographic area. This history reflects a customer growth rate of around 2 to 3 percent per year, with a gradual decrease in the growth rate. The graph of the past history is relatively smooth, and an extrapolation of this curve can be made for future years with a good degree of confidence.

Similar projections can be made for each classification of customer, with the possible exception of the large industrial customers. Each individual gas company usually has specific information on the large industrial customers they already serve, as well as any prospective customers that they are working with. This is a much more accurate approach than the historical growth pattern for this class of customer.

Monthly Number of Customers

Although the annual average number of customers usually gives a smooth historical curve, this is not true of the monthly number of customers. Figure 2-2 shows this for a typical residential customer group. Table 2-2 shows this same information in tabular form. Part of this variation is due to the normal, steady growth in the number of customers. This effect results in the December value being consistently higher than the

14 Underground Gas Storage Facilities

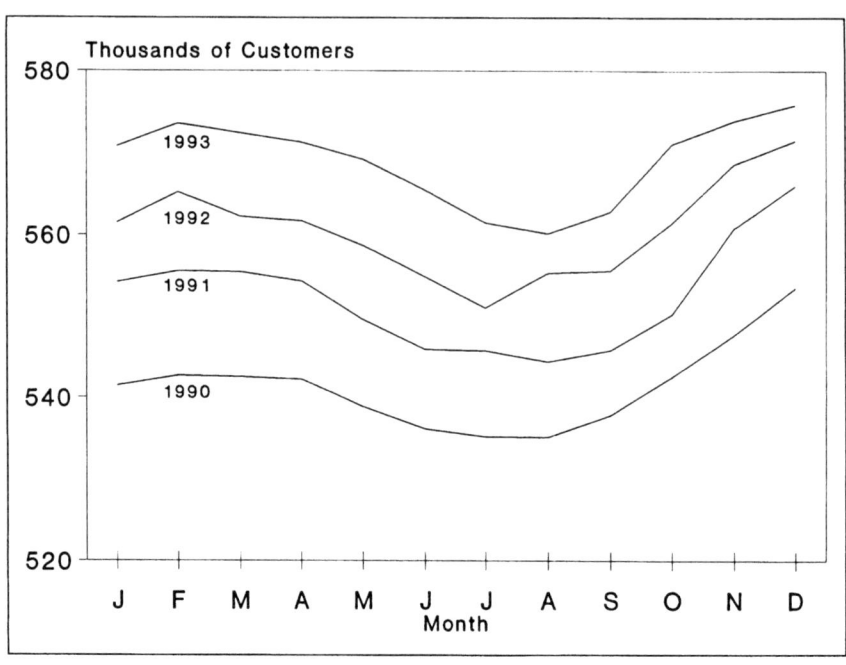

Figure 2-2. Monthly variation in number of residential customers.

TABLE 2-2
Monthly Number of Residential Customers

Month	Number of Customers			
	1990	1991	1992	1993
January	541,450	554,130	561,506	570,810
February	542,662	555,510	565,172	573,560
March	542,486	555,392	562,196	572,398
April	542,200	554,270	561,690	571,266
May	538,806	549,546	558,652	569,204
June	536,114	545,906	554,840	565,482
July	535,170	545,746	551,006	561,494
August	535,124	544,374	555,280	560,176
September	537,812	545,784	555,546	562,818
October	542,496	550,238	561,462	571,080
November	547,674	560,768	568,628	573,842
December	553,466	565,966	571,472	575,794
Annual Average	541,288	552,303	560,621	568,994

January value. The major factor in the variation in the monthly number of customers, however, is that many customers will have their gas service disconnected when the space-heating season is over and have it re-connected in the autumn. Where space heating is the only gas utilization, it is less costly to pay the re-connection charge and eliminate the minimum bill each month. The monthly variation in the number of customers may be represented by a factor. This factor is obtained by dividing the number of customers for each month by the average annual number of customers for that year. These factors for an individual month may be averaged for several years to yield a factor for each month to be used for design calculations. Table 2-3 illustrates this.

Small commercial and small industrial customers have especially wide fluctuations in the monthly number of customers. Many of these cus-

TABLE 2-3
Monthly Number of Residential Customers as a Multiple of the Average Annual Number

Month	Number of Customers Factor				
	1990	1991	1992	1993	Average
January	1.00030	1.00331	1.00158	1.00319	1.00209
February	1.00254	1.00581	1.00812	1.00803	1.00612
March	1.00221	1.00559	1.00281	1.00598	1.00415
April	1.00168	1.00356	1.00191	1.00399	1.00279
May	0.99541	0.99501	0.99649	1.00037	0.99682
June	0.99044	0.98842	0.98969	0.99383	0.99059
July	0.98870	0.98813	0.98285	0.98682	0.98662
August	0.98861	0.98564	0.99047	0.98450	0.98731
September	0.99358	0.98820	0.99095	0.98915	0.99047
October	1.00223	0.99626	1.00150	1.00367	1.00092
November	1.01180	1.01533	1.01428	1.00852	1.01248
December	1.02250	1.02474	1.01936	1.01195	1.01964

tomers have no gas usage except for space heating. Therefore a high percentage of these customers will have their gas service disconnected during the warmer months.

The large commercial, medium industrial, and large industrial customers usually show no patterned variation in the monthly number of customers similar to that shown in Figure 2-2. These customers have enough process, cooking, or other gas needs year round that they cannot forego their gas service during the summer months. These customers present a constant customer base month to month.

In order to accommodate this monthly variation in the number of residential, small commercial, and small industrial customers, equation 2-1 must be modified:

$$\text{Gas Usage} = N \times M \times U \times DD \times D \qquad (2\text{-}2)$$

N = annual average number of customers
M = monthly number of customers factor
U = usage per customer in Mscf per degree day per day
DD = average number of degree days per day for the period
D = number of days in the time period

This equation would give the gas usage for the residential customers for any time period within a given month. In order to develop the gas usage for a year, the usage must be calculated for each month.

Usage Per Customer

For residential gas customers, there is usually a base load that is independent of atmospheric temperature. This consists of cooking, water heating, clothes drying, and miscellaneous uses such as gas lights and gas grills. In addition there is a space-heating load that is dependent on the atmospheric temperature.

It has been generally conceded by gas companies that the space-heating load starts when the temperature drops below 65°F. This has given rise to the degree day concept. A degree day has been defined as a day in which the average temperature is 1 degree below 65°F. Thus a day with an average temperature of 43°F would have 22 degree days per day. The summation of the individual degree days in a month would give degree days per month.

Heating loads in a given area are generally correlated with degree days, although there has been some recent work [1] to correlate the space-heating usage without degree days. The basic principle is still contained, however.

The usual procedure to develop degree day correlations is to plot average daily gas usage per customer for a single type of customer versus degree days per day. Figure 2-3 shows such a plot. The data includes only residential customers, but the graph includes the entire time period of the heating season, October through April. Experience has shown that better correlations can be obtained if plots are made of the individual months.

Load Forecasting 17

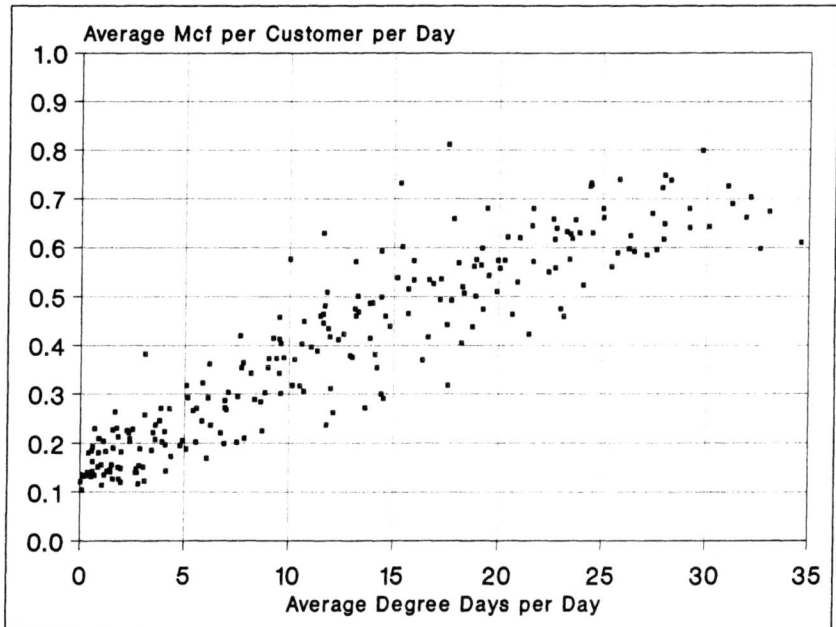

Figure 2-3. Residential customer gas consumption by degree days.

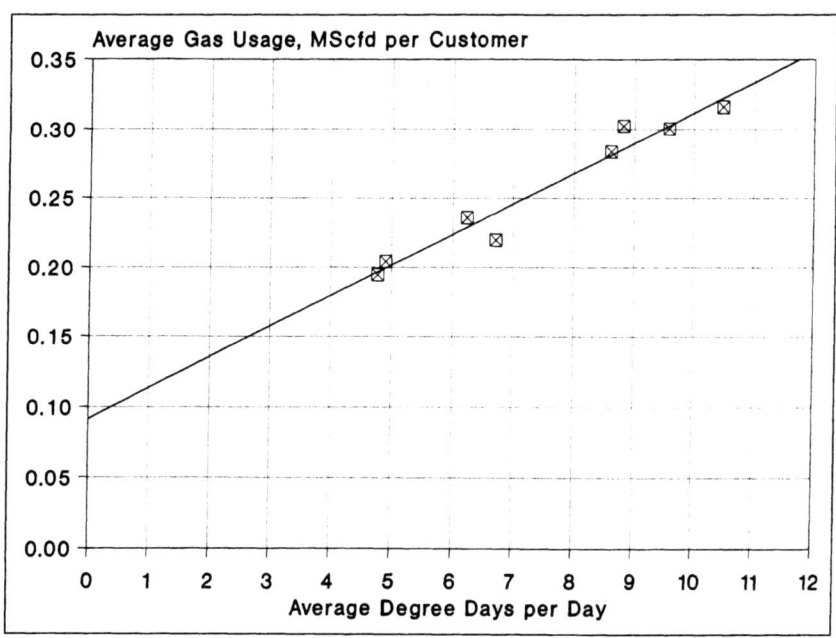

Figure 2-4. Residential customer gas consumption, November.

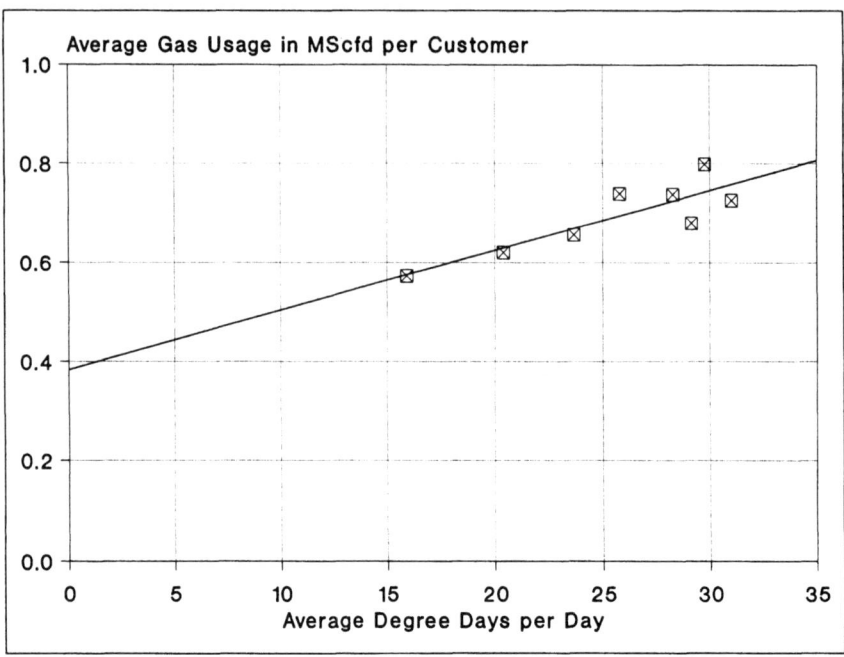

Figure 2-5. Residential customer gas consumption, January.

Figure 2-4 shows a graph of customer usage versus degree days per day for November. Figure 2-5 shows a similar plot for January. A significant difference can be noted between the two. The intercept at zero degree days for the month of November is much lower than the intercept for January. This says that with no heating load, the gas usage in January is higher than in November. One reason for this may be the water-heating load. The ground is colder in January than it is in November, so the water entering the water heater is colder in January and requires more gas to raise it to the thermostat temperature. In addition, the cold ground temperature in January may cause houses built on concrete slabs to require some heat even though the atmospheric temperature is above 65 degrees.

In addition, the slope of the correlation line for November in Figure 2-4 is steeper than the line for January in Figure 2-5. One possible explanation for this is that not everyone turns on their space heater at 65°F at the beginning of the heating season. Thus the November correlation may represent not only the increased consumption by a given number of customers due to colder weather but also an increased number of customers who are turning their heating systems on. In January it is presumed that the vast majority of people already have their systems on, so that this effect is not seen.

Similar correlations can be developed for the residential customers for the other months in the heating season. For the months that are not in the heating season, the correlation would simply be the average base load per customer for each of the months.

The correlations just discussed are for residential customers. Similar correlations can be made for small commercial, large commercial, small industrial, and medium industrial customers. Generally, large industrial customers are not susceptible to this treatment. Large industrial customers usually have a budget that projects their production schedules. Gas companies usually interview these companies on an individual basis in order to develop their projected gas consumption.

Non-Consumer Gas Requirements

Company-used Gas. Gas companies are themselves customers for gas usage. These uses may include compressor fuel, fuel for gathering line heaters, space heating and other needs in company offices, and similar uses. This category should also include lost and unaccounted-for gas. All of these uses should be included in the company-used category because they affect the gas balance and influence storage needs.

Transported and Exchange Gas. When transported and exchange gas are contemporary and balanced in their receipt and delivery, they usually do not affect storage because the supply equals the requirement. They may, however, affect transmission system capacity and should be included in the gas balance. Many times there is a time difference between the receipt of the exchange and transport gas and the delivery of it. This time difference may vary from a few hours to several months. This time offset can cause a need for storage capacity.

Total Requirements

Once the customer use correlations have been developed for each type of customer, the gas load can be forecast for each type of customer for each month of the year using the number of degree days for each month for a normal year. To these forecasts can be added the monthly estimates for large industrials, company used, and transported and exchange gas. This gives the total requirements by month for a normal weather year, and this information can serve as a basis for financial forecasting. Rate schedules may be incorporated into this calculation to convert the gas requirements into revenue. In addition, percentage multipliers may be incorporated into Equation 2-2 to show the effect of a winter that is 10 percent colder (or 15 percent warmer) than normal. This can be a very powerful tool for financial budgeting and forecasting.

In many cases it is desirable to forecast the gas requirements by day for a year. This can be done by using the historic weather data that has been processed as described in Chapter 1 and equation 2-2. For gas requirements where only monthly data is available and where the daily load is relatively constant within a given month, the monthly requirement may be divided by the number of days in a month to give the daily values. Figure 2-6 shows the result of this total requirements calculation for each day of the year. This figure shows the large variation in gas usage that is caused by the daily fluctuations in average daily temperature. Of particular interest in Figure 2-6 is the dramatic dip in requirements near the end of the year. The historic data for this particular weather station showed that there is normally a day that has an average temperature of 69°F, and this temperature would probably occur on December 19.

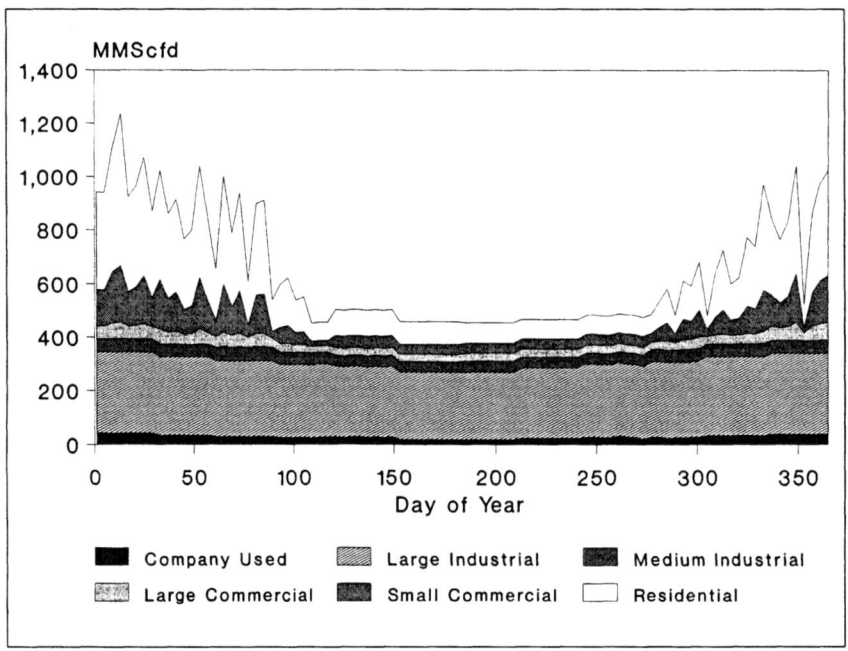

Figure 2-6. Total requirements—chronological data presentation.

Daily Weather Forecasting

The method described above gives a daily forecast of gas requirements that is based on averages and is sufficiently accurate for budgeting and storage design work. There are times, however, when a gas company needs to know with better accuracy on a short-term basis what the sendout will be. This is particularly true of large distribution companies who have a forecast of temperature and need to know the magnitude of the gas requirements for tomorrow and the day after so that they can manage their supplies and possibly schedule interruptions for their interruptible customers if necessary.

Figure 2-5 correlates historical monthly gas usage data for January, and the data shown represents what can be considered a good fit for this purpose. Some of the data in Figure 2-5, however, varies from the linear curve fit by as much as 6 percent. Many companies would regard this as unsatisfactory for short-term forecasting purposes. There are ways to improve the accuracy of this forecast for specialized cases.

Figure 2-7 shows the gas sendout by hours for a gas company that has a mix of residential, commercial, and industrial customers. The shape of

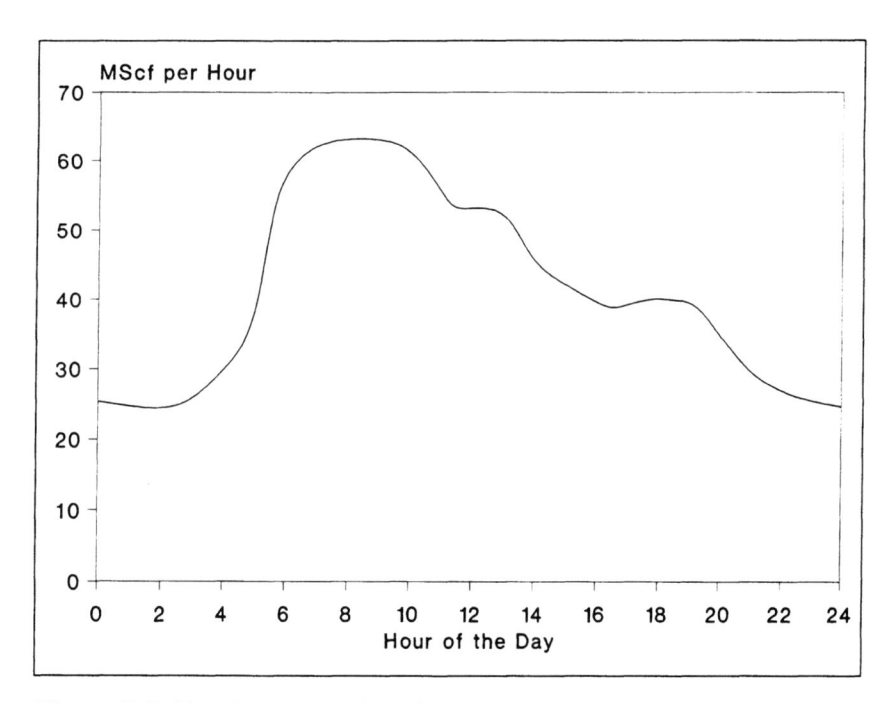

Figure 2-7. Hourly gas sendout for gas company with typical distribution customer mix.

the curve will be different for different customer mixes, but Figure 2-7 represents a typical, if somewhat hypothetical, situation with a given proportion of residential, commercial, and industrial customers for a cold winter day.

During the night hours the space-heating thermostats are turned down, the commercial loads are low, and the total demand for gas is relatively low. At about 5:30 A.M. residences begin to turn up the thermostats, commercial establishments start up, and the big load of the day comes on. Cooking is added to this, and the load peaks at about 8:00 to 9:00 A.M. As the space-heating thermostats catch up, the load drops off around 10:00 A.M. This is accelerated by the fact that in many homes the residents leave to go to work, school, or other activities, and the thermostats are turned down.

Around 11:30 A.M. the lunch cooking load can be seen coming on, and this lasts until about 1:30 P.M. As the day warms up, the gas load continues to drop until about 4:30 P.M., when residents are returning home. In addition, the supper time cooking load starts up and lasts until around 7:30 P.M. As residents go to bed, the load continues to decrease. This pattern remains essentially the same for any customer grouping that has a large residential component, although the shape of the curve may vary somewhat with the proportion of commercial and industrial customers.

Many gas companies have attempted to correlate this total daily sendout with the average daily temperature and have had fair success. There are reasons why this type of correlation does not always produce satisfactory results. The average daily temperature is a useful tool, but it is the average of the high temperature and the low temperature for the day and gives no information on the temperature pattern or profile for that day. Consider the examples shown in Table 2-4. Day 1 and day 2 have identical high and low temperatures, and this gives identical average daily temperatures. The weighted average temperatures for the two days, however, are 2 degrees apart. Thus the high-low temperature average does not adequately define the temperature pattern for accurate forecasting.

Day 2 and day 3 show another interesting reason for the lack of correlation. These two days have the same high-low temperature average and the same hourly weighted temperature average. Day 2 represents a rising temperature pattern, while day 3 depicts a falling temperature pattern. If gas loads were consistent throughout the day, these two days should have identical gas loads. Because of the heavy gas usage in the early daylight hours, however, day 2 with its low temperatures during this period will use much more gas during this time period than day 3. This will make the total gas usage for the day considerably greater for day 2 than for day 3.

TABLE 2-4
Examples of Daily Temperature Patterns

Hour of Day	Day 1	Day 2	Day 3
0	14	14	30
2	14	15	29
4	15	16	28
6	15	17	27
8	18	18	26
10	20	20	24
12	22	22	22
14	26	24	20
16	30	26	18
18	26	27	17
20	24	28	16
22	20	29	15
24	16	30	14
High Temperature	30	30	30
Low Temperature	14	14	14
Hi-Lo Average	22.0	22.0	22.0
Weighted Average	20.0	22.0	22.0

In order to eliminate this difficulty, the day can be divided into four six-hour periods. The gas sendout for each period can be correlated against the average temperature for that period. This refinement gives a substantial improvement in the correlation. It is then necessary to get temperature forecasts for each six-hour period in order to forecast the sendout. These six-hour forecasts can usually be obtained from a private weather service.

Other weather factors have been tried as correlation factors. These include percent cloud cover, wind direction and velocity, hours of sunshine, and others. The effect of these, however, is usually so small that they are not good correlation factors. A factor that is worth mentioning is day of the week. On weekends the residential load is usually greater than on weekdays because more people are at home and use more gas for space heating. Conversely, more businesses are closed and use less gas. Monday morning usually brings a large gas usage for commercial and industrial customers as they start up businesses that have been shut down over the weekend.

The forecasting methods described here are methods that have been used successfully in the past. They are not the only methods available or in use. There has been a good deal of interest in load forecasting recently, and some new work has been done in this field [1,2].

References

1. Deming, Matthew D. *A Theory of Weather-Sensitive Energy Demand.* Paper presented at the 25th Annual Meeting of the Pipeline Simulation Interest Group, October 14–15, 1993.
2. Aronov, Edward. *Development of the Adaptive Model for Forecasting Energy.* Unpublished paper.

CHAPTER 3

Load Curves

Chapter 2 described the method for making a daily gas load forecast for each day of the year based on historical weather data that had been processed according to the methods described in Chapter 1. The result of this endeavor is shown in Figure 2-6, where the forecast for each day of the year is shown for each type of customer in chronological sequence. While Figure 2-6 is interesting, it is difficult to work with to develop quantitative information.

Requirements Load Curves

When the data in Figure 2-6 is sorted as to magnitude of total gas demand by day and the result is displayed in descending order, the result is shown in Figure 3-1. The y-axis remains the same, but the x-axis

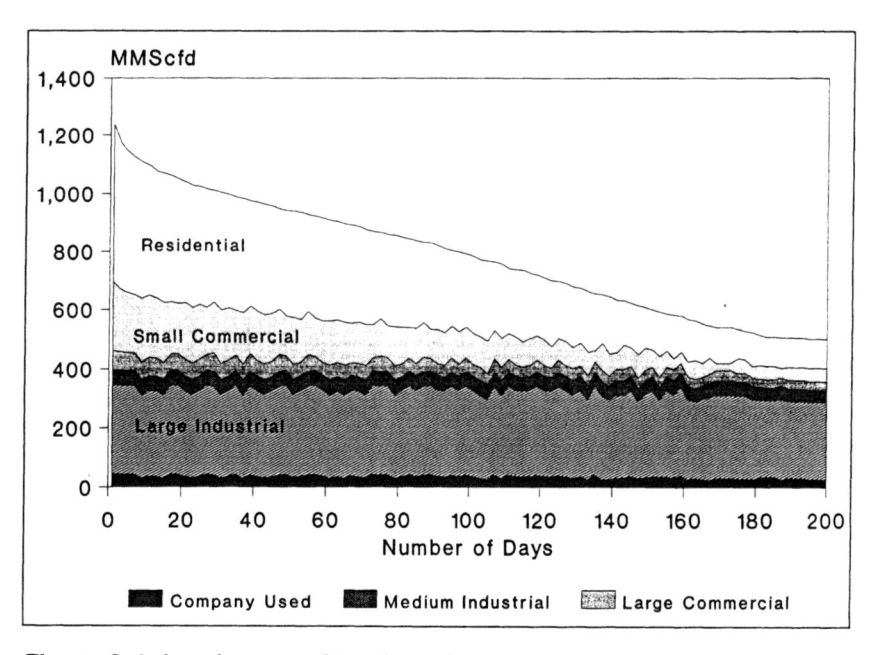

Figure 3-1. Load curve of total requirements.

26 *Underground Gas Storage Facilities*

changes. The *x*-axis is no longer days of the year. It is now the number of days that the gas load is the value shown at that intercept or greater. For example, at an *x*-axis value of 40, there are 40 days that have a total gas load of 985 MMScf per day or greater. This figure is called a load curve.

An examination of Figure 3-1 shows little seasonal variation in the gas load for the large and medium industrial and the large commercial loads. This reflects the large proportion of process gas that makes up their total load. Some month-to-month variation shows up as a small ripple effect. The large seasonal variation is in the residential and, to a smaller extent, the small commercial loads. Both of these types of customers have significant space-heating loads.

Requirements and Supply Load Curves

Figure 3-1 shows only the requirements portion of the gas balance. The other portion is the supply portion. The total requirements portion of Figure 3-1 may be used to see how the gas supply matches this requirement. Figure 3-2 does this. The total requirements in Figure 3-2 are identical to those in Figure 3-1. This particular hypothetical gas company has three sources of gas supply.

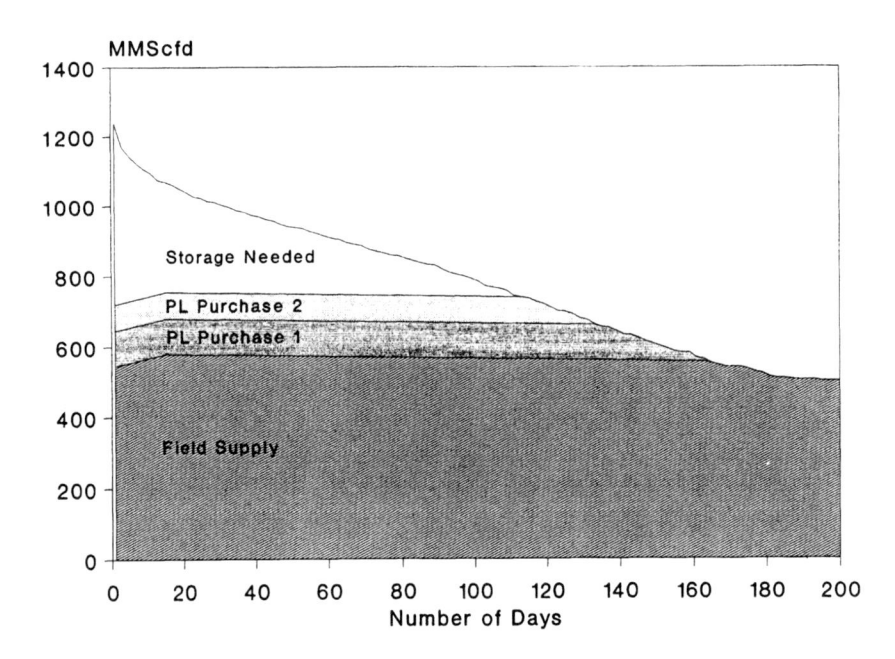

Figure 3-2. Load curve of total requirements and supply.

The major source of supply for this particular company is a field supply consisting of a large number of onshore gas wells and possibly several offshore wells. The characteristics of these wells are similar to almost all gas wells and are reflected to some extent in the figure. The field supply shows a peak at about 15 days and drops off significantly to the left of day 15. On very cold days it is assumed that a portion of the field supply will not be available due to the cold weather. This decrease in availability may be due to the wells freezing up, the gathering lines getting plugged with hydrates, icy roads preventing switchers from getting to the wells to turn them on, or other cold-weather causes. For this particular field supply it is assumed that 6 percent of the total field supply will be lost on the coldest day of the year, and that this decrease will be linear from day 15 to day 1 on the load curve.

Figure 3-2 also shows that the field supply decreases to the right of day 15. This decrease is caused by the normal decline in well deliverability with production. For this particular case, it is assumed that this decline is 10 percent per year. While some new production will offset this decline on an annual basis, it is doubtful that these new wells can be counted on to come on during the coldest part of the year. Thus the peak field supply occurs at day 15 of the load curve.

Two pipeline purchases make up the remainder of the supply. The nature of pipeline purchases may vary widely according to the terms of the purchase contract. Some contracts may require that an annual volume be taken or a penalty is imposed. A common restraint is that no more than a contract maximum amount of gas may be taken on any day. For purposes of this example, it is assumed that the pipeline purchases have a contract maximum daily take, and no other restrictions other than relative price apply. It is assumed that pipeline purchase two has more expensive gas than pipeline purchase one, and that both pipeline purchases are more expensive than the field gas.

It can be seen from Figure 3-2 that at day 110 and beyond, the supplies are adequate to furnish all of the needs of the customers. There will be 109 days, however, when the normal supply will be inadequate. For these 109 days some form of supplementary gas supply will be needed. The decision as to whether to purchase additional supply or to use underground storage is an economic one. A gas purchase that supplies gas for only the coldest days of the year can be very expensive, because that is the time when all gas companies want to have additional gas. If gas is available for purchase at that time, it will probably come from underground storage from another company. For purposes of this example it is assumed that the purchase of an additional gas supply would be economically unwise, and that gas from some underground storage must be used. Whether this underground storage service is purchased or the underground storage is developed within the company is an economic choice.

An examination of Figure 3-2 shows that the total amount of the requirements that the normal supply is not capable of handling during the colder days is 20,625 MMScf. On the coldest day the normal supply is deficient by 517 MMScf per day. Any storage service must be capable of handling these deficiencies and more. The load curve shown in Figure 3-2 was developed for a normal winter. Since some winters will be, statistically, colder than normal, provisions must be made for these cases. In order to provide for this, a load curve must be made up with data from a colder than normal winter. The decision of how much colder than normal to provide for is something the planner must work out. The Weather Bureau historical data will be of help in determining what is a 10-year, a 30-year, or a 50-year extreme.

Supply and Storage Load Curves

It would probably be difficult to find a single storage reservoir that could handle the storage needs in Figure 3-2 economically. A better choice might be to attempt to find two or more reservoirs to do the job. Figure 3-3 shows one manner in which the storage needs may be satisfied using three storages.

Table 3-1 shows the minimum required capacity and deliverability for each of the three storages in Figure 3-3. It can be seen that the three storages have quite different characteristics. Although the characteristics of storages will be discussed in depth in a later chapter, it might be well to touch on this here. Storage 1 has a relatively large storage capacity and a

Table 3-1
Minimum Storage Requirements for Hypothetical Gas Company

	Capacity MMcf	Deliverability MMcfd
Storage 1	9,750	100
Storage 2	8,907	175
Storage 3	1,968	242
Total	20,625	517

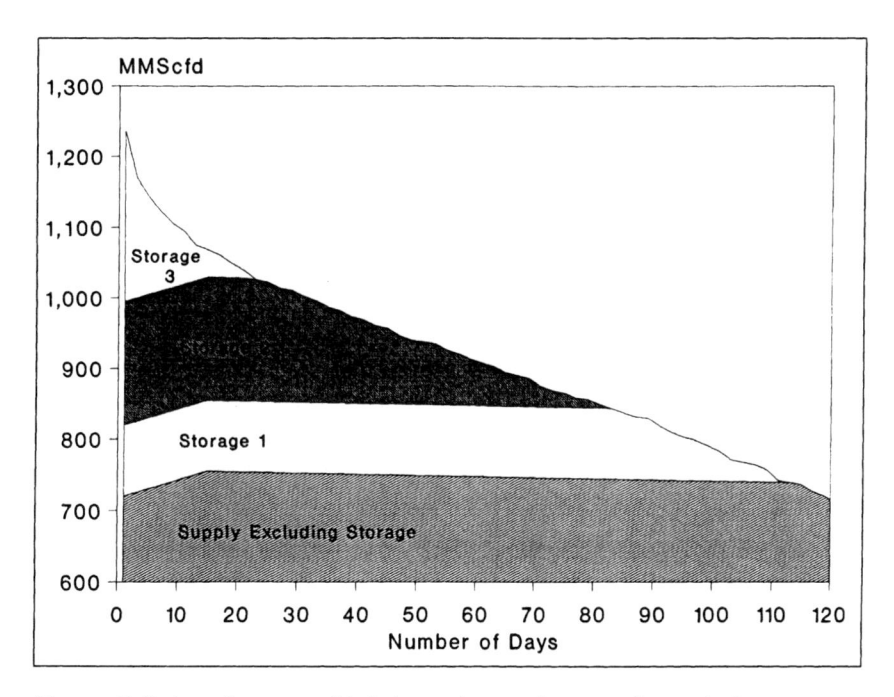

Figure 3-3. Load curve of total requirements, supply, and storage.

relatively low deliverability. By contrast, storage 3 has a relatively low storage capacity and a high deliverability. Storage 2 is in between the two. Storage 1 is a base type of storage that would come on as soon as storage is required and would remain producing throughout the bulk of the winter season. Storage 3 would not be used until the coldest days of the year, and would be turned on and off on a daily basis as needed. Because of its limited capacity, storage 3 must be used judiciously and turned on only when peaking gas is needed.

Since storage is a temporary supply, any gas that is removed from storage must be replaced. This replacement should occur during the same calendar year following the winter of withdrawal. If the replacement is not done in this manner, there will not be storage gas available to withdraw the following winter. Gas injection into storage can occur any time that the total supply exceeds the total requirements. Figure 3-4 shows how this injection-withdrawal balance might be achieved for the company being used for illustration. The withdrawals occur in varying amounts through day 109. Following this is a period of about 49 days when there is no injection or withdrawal. These days may actually be interspersed throughout the winter season. They are warmer days occurring between cold weather periods, or they may be colder days occurring between warmer weather periods.

At about day 158 injection begins at low rates. As the customer requirements decrease due to warmer weather or other causes, the injection rate increases somewhat. In this particular case, the beginning of injection on day 158 roughly corresponds to the time when the gas supply from the two pipeline purchases is no longer needed. This is fortunate, in that it is probably more economical to use the lower-priced field gas to replenish storage than to use the higher-priced pipeline purchases. As mentioned earlier, the field supply will decline in deliverability as the wells are produced. If the loss of deliverability is not replenished by the addition of new field supplies, it may be necessary to use some gas purchased from the pipelines near the end of the injection season in order to meet the required injection rate.

It should be remembered that Figure 3-4 is an idealized schedule. During the colder weather, the withdrawals will occur as needed because the gas company has little choice. The main variables during this period are the accuracy of the load forecasts and the departures of the weather from

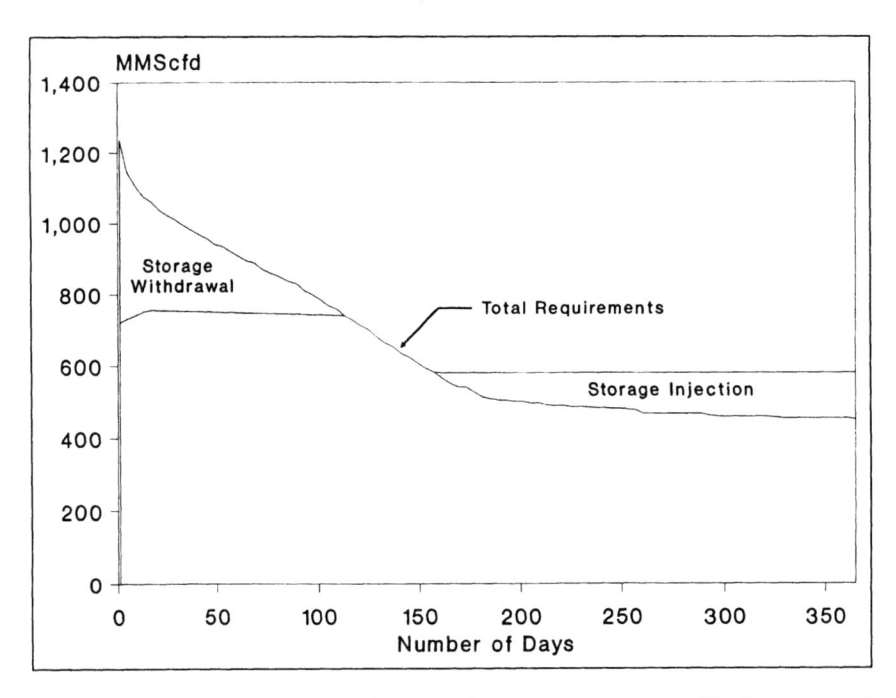

Figure 3-4. Load curve showing requirements, storage withdrawal, and storage injection.

the historic normals. During the injection season, however, the gas company has some latitude as to timing. If maintenance is scheduled on the storage facility equipment, the injection schedule may be interrupted for 20 or 30 days. Conversely, some injection may occur during the warmer days of the winter. There are cases of withdrawing from storage for part of a day and injecting into storage for another part of the same day.

As mentioned earlier, the figures shown here were developed for normal winters based on historic data from the Weather Bureau. It is rare to have a normal winter. The main question is how far will be the divergence from normal and what will be the timing. For this reason it is essential to prudent planning and management that the non-normal cases be examined. The warmer than normal winters mean that the storages will not be fully exercised. The petroleum engineer should be consulted to determine if any corrective measures are needed for the reservoir in this case. For colder than normal winters, it is vital that the storages are sized adequately. It may be desirable to examine the effect of winters that are 20-percent warmer than normal to winters that are 30-percent colder than normal. This examination should be made from the viewpoint of both the storage facilities and the financial impact on the gas company.

CHAPTER 4

Gas Laws

Most substances are relatively easy to specify as to quantity. A pound of a solid material is definitive. A gallon of a liquid is almost definitive, with only a minor variation due to temperature and pressure. A cubic foot of gas, however, is not at all definitive. This is because gases are compressible. They are greatly affected by temperature and pressure.

Pressure Measurement

Pressure may be measured by several different instruments. Some of these are pressure gauges, liquid-filled manometers, and the various types of pressure transducers. All of these devices have one common characteristic—they are actually measuring a differential pressure. An example is the liquid-filled manometer. One leg of the manometer is connected to the pressure being measured. The other leg is open to the atmosphere (or to some other pressure). When one leg is open to the atmosphere, the device is measuring the differential pressure between the pressure vessel and the atmosphere.

The pressure gauge acts in a similar manner. This device contains a coiled tube, called a Bernoulli coil. As pressure is introduced into the coil, the coil unwinds and moves a pointer that indicates pressure. The coil is actually moving against the atmospheric pressure that is on the outside of the coil. Therefore, the measurement made is differential pressure between the pressure being measured and the atmospheric pressure. A pressure gauge could be made that would measure the absolute pressure in a vessel. This would require that the case of the gauge be evacuated to a high vacuum. The gauge would then be measuring the difference between the pressure vessel and zero, and the result would be the absolute pressure. Such a gauge would be expensive to manufacture and practically impossible to maintain.

Some pressure gauges have been made with no zero point at the bottom of the scale. Instead, the value of the atmospheric pressure has replaced the zero point and all the scale was elevated by this amount. The problem with this arrangement is that the atmospheric pressure varies from day to day and from place to place, and this variation can be significant. Because of this, practically all pressure gauges have a zero point and measure differential pressure. This pressure measurement is called gauge pressure, as opposed to absolute pressure.

Temperature Measurement

There are several temperature scales. These include Celsius, Fahrenheit, and Ranier, which beer makers use. When these scales were developed, there was little thought given to the total span of temperature. The scales were patterned to provide a convenience for everyday living. Although often obscure, there was some logic to the scale development. The Celsius scale took the boiling point of water as one point and the freezing point of water as another and divided the distance between them into 100 degrees. The logic behind the Fahrenheit scale is more vague, and different theories exist on why the values were chosen.

The fact that all the scales had negative values indicated that the developers and users early on recognized that there was a significant area not covered by the positive scale values. Further scientific work revealed that there is an absolute lower limit of temperature. This lower limit has been determined to be −460°F and −273°C. This is a theoretical value that has never been reached. There is some thought that although this lower limit is well defined, it is as unachievable as the upper limit of infinity. When dealing with material properties that are temperature sensitive, it is sometimes necessary to recognize this lower temperature limit in order to quantify this temperature-sensitive relationship. As a result of this new information, two additional temperature scales were developed, the Rankine and the Kelvin. These are sometimes referred to as the absolute Fahrenheit and the absolute Celsius scales. The Rankine scale has the same degree size as the Fahrenheit, but its zero point is equal to −460°F. The Kelvin scale relates to the Celsius scale in a similar manner.

The Effect of Pressure on Gas Volume

When pressure is applied to a given quantity of gas, the volume shrinks, which is illustrated in Figure 4-1. In the first diagram, gas is confined in a cylinder with a sliding piston. Sufficient weight is on the piston to produce a pressure of 35 psig and the gas volume is 10 cubic feet. The second diagram shows the same vessel when more weight has been added to the piston. The pressure is now 85 psig and the volume is 5 cubic feet.

34 Underground Gas Storage Facilities

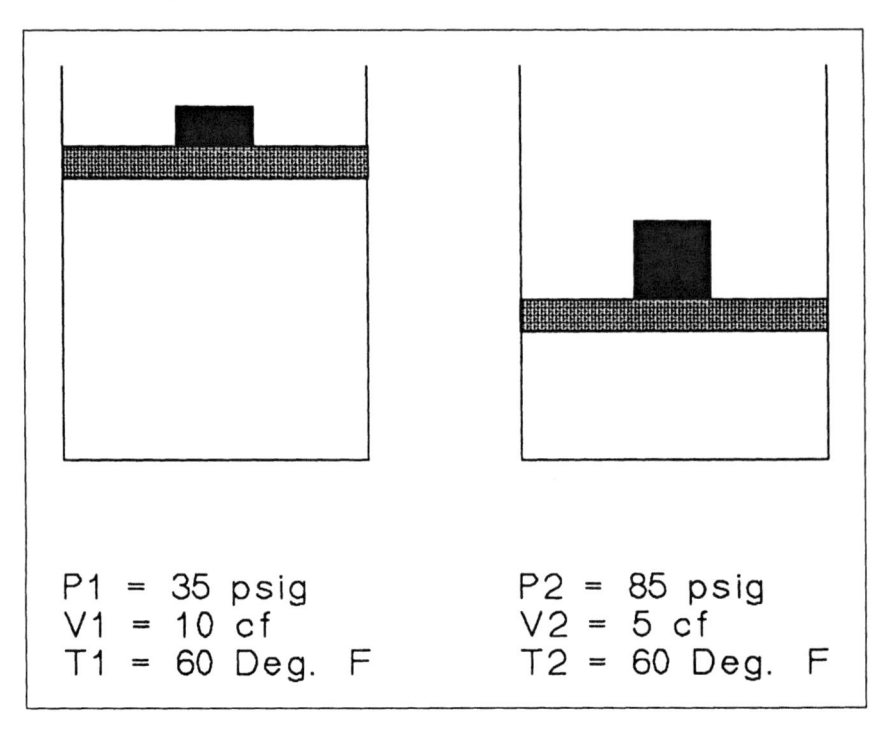

```
P1 = 35 psig         P2 = 85 psig
V1 = 10 cf           V2 = 5 cf
T1 = 60 Deg. F       T2 = 60 Deg. F
```

Figure 4-1. The effect of pressure on gas volume at constant tempera-

Boyle's law states that if a gas is at constant temperature, its volume is inversely proportional to the pressure, or

$$P_1 \times V_1 = P_2 \times V_2 \tag{4-1}$$

A casual inspection of Figure 4-1 would tend to indicate that the gas in the cylinder does not follow Boyle's law. It must be remembered, however, that the pressures of 35 psig and 85 psig are gauge pressures or differential pressures with the atmosphere. When the atmospheric pressure (approximately 15 psia) is added to the gauge pressures, the result is 50 psia and 100 psia. The gas volumes and the absolute pressures are then inversely proportional.

The Effect of Temperature on Gas Volume

When the temperature of a quantity of gas is raised, the volume of the gas increases, which is illustrated in Figure 4-2. In the first diagram the gas temperature is 60°F and the gas volume is 5 cubic feet. The second

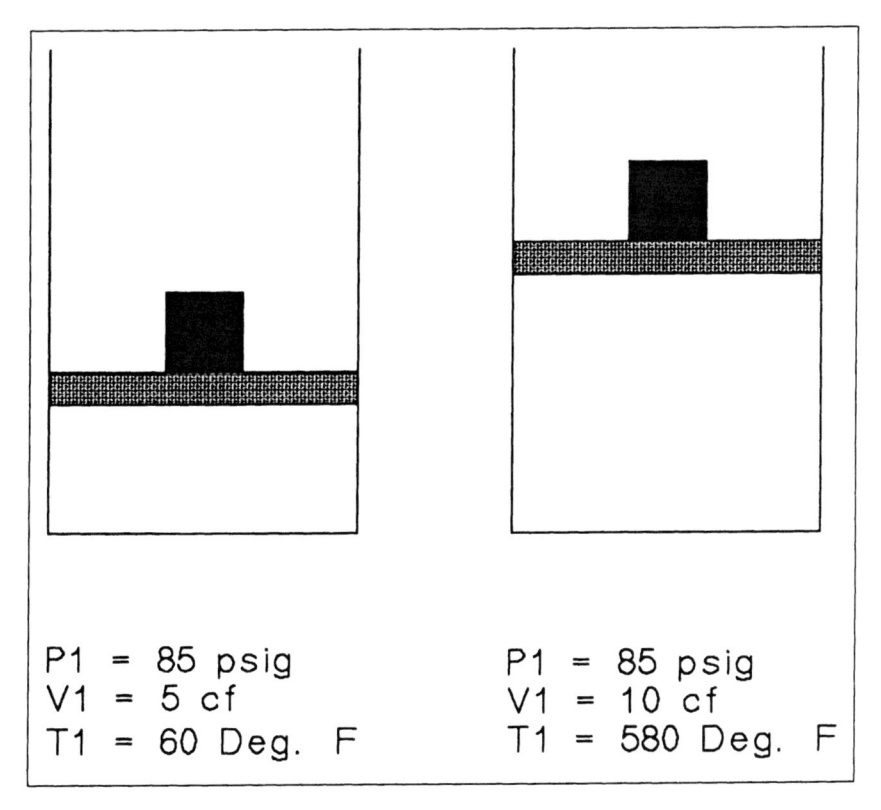

```
P1 = 85 psig        P1 = 85 psig
V1 = 5 cf           V1 = 10 cf
T1 = 60 Deg. F      T1 = 580 Deg. F
```

Figure 4-2. The effect of temperature on gas volume at constant pressure

diagram shows that when the temperature is raised to 580°F, the gas volume doubles to 10 cubic feet, the pressure remaining constant.

Charles' law states that the volume of a gas is directly proportional to the temperature of the gas when the pressure remains constant or:

$$\frac{V_1}{T_1} = \frac{V_2}{T_2} \tag{4-2}$$

Again, the diagrams in Figure 4-2 do not seem to support this law until it is remembered that the temperatures must be on an absolute scale. To convert the Fahrenheit temperatures to absolute temperatures (degrees Rankine), 460 must be added to the Fahrenheit values. When this is done, the two temperatures become 520 and 1040° Rankine, respectively. With these values, Equation 4-2 does hold for Figure 4-2.

Ideal Gas Law

The laws of Boyle and Charles may be combined into a single equation:

$$\frac{P_1 \times V_1}{T_1} = \frac{P_2 \times V_2}{T_2} \qquad (4\text{-}3)$$

This is called the *ideal gas law*. This law also may be expressed in a different form by making the following derivation where the "s" subscript denotes standard conditions:

$$\frac{P \times V}{T} = \frac{P_s \times V_s}{T_s} \qquad (4\text{-}4)$$

$$V_s = n \times MV \qquad (4\text{-}5)$$

n = number of mols of gas
MV = standard volume occupied by one mol of gas

The term "n" is expressed in mols. A mol is a convenient method of specifying quantities of gases. A mol of any substance is the molecular weight of the substance expressed in weight units. As an example, oxygen has a molecular weight of 32. A pound mol (or pound molecular weight) of oxygen would be 32 pounds of oxygen. Similarly, a gram mole (or gram molecular weight) of oxygen would be 32 grams of the gas. A pound mol of any gas always occupies the same volume at standard conditions. This volume is 378.58 cubic feet. An informal standard is that pound molecular weights are referred to as "mols" while gram molecular weights are called "moles," although this convention is not at all adhered to rigidly.

Substituting Equation 4-5 into Equation 4-4:

$$\frac{PV}{T} = \frac{P_s \times MV \times n}{T_s} \qquad (4\text{-}6)$$

$$R = \frac{P_s \times MV}{T_s} \qquad (4\text{-}7)$$

Rearranging:

$$P \times V = n \times R \times T \tag{4-8}$$

P = pressure
V = volume
T = temperature in absolute units
n = number of mols of gas
R = universal gas constant

The gas constant is a number that is used to account for the units that are used for P, V, T, and n. Table 4-1 lists the value of the gas constant R for several sets of dimensional units.

Table 4-1
Values of Gas Constant R For Various Dimension Units

P	V	T	n	R
Atm.	Cu. Cm.	Deg. K	Gr. Mol	82.06
psi	Cu. Ft.	Deg. R	# Mol	10.73
psf	Cu. Ft.	Deg. R	# Mol	1544
Atm.	Cu. Ft.	Deg. R	# Mol	0.73
Atm.	Cu. Ft.	Deg. K	# Mol	1.315
In. Hg.	Cu. Ft.	Deg. R	# Mol	21.85

Compressibility Factors

The ideal gas laws work well at relatively low pressures and relatively high temperatures. When the pressure and temperature depart from these ranges, significant error can result from the use of the ideal gas laws. At high pressures and low temperatures, for example, a gas will occupy a smaller volume than is predicted by the ideal gas law. One hypothesis has been advanced that as the gas molecules are crowded together the gravitational attraction between molecules becomes a factor and this attraction causes the gas volume to be less than calculated. At very high pressures, the reverse is true; the gas occupies a greater volume than is computed by the ideal gas law. One explanation for this is that as the gas molecules become crowded very close together, the physical size of the molecule begins to become a factor and the gas begins to become slightly incompressible. Regardless of the reasons, the ideal gas law does not accurately represent the behavior of gas at high pressures.

In order to study this problem, much research work has been done. The result of this research work is a term called the *compressibility factor*. This is also sometimes called the *supercompressibility factor*. The American Gas Association has sponsored research that defines these factors for all of the conditions to which natural gas is normally exposed. The factors have been found to be affected by temperature, pressure, gas specific gravity, and gas composition, particularly the inert content of the gas. This information is in the form of tables of values as well as the equations for calculating the factor values from the gas properties and conditions.

When the compressibility factor is incorporated into the ideal gas law equation, the result is:

$$P \times V = n \times Z \times R \times T \tag{4-9}$$

where Z is the compressibility factor determined by whatever method is appropriate.

Approximations

The compressibility factor equations developed by the American Gas Association and others are intended to be used with natural gases with a wide range of properties and exposed to a wide range of conditions. By necessity, these equations are quite sophisticated, lengthy, and require considerable computer effort. In a given storage reservoir, there is little variation in temperature and the range of pressure variations is relatively small. Furthermore, the composition of the gas will not vary dramatically with time. For this type of situation an approximate equation for compressibility factors is appropriate.

The total gas in place in the reservoir should be a straight line function of the pressure divided by the compressibility factor, or:

$$GIP = a + b \times \frac{P}{Z} \tag{4-10}$$

GIP = total gas in place
P = pressure in psia
Z = compressibility factor
a = constant
b = constant

Gas Laws

The compressibility factor can be computed in an approximate manner for the limitations imposed above by the following form of equation:

$$Z = c + d \times P \tag{4-11}$$

c = constant
d = constant

By combining these two equations the total gas in place in the reservoir may be calculated as a function of the pressure:

$$GIP = a + \frac{b \times P}{c + d \times P} \tag{4-12}$$

This equation can be used to estimate total gas in place in the reservoir with reasonable accuracy. Appendix A shows how to evaluate the constants a, b, c, and d for a specific reservoir. It should be emphasized that Equations 4-11 and 4-12 should be used only where there is a limited range of variation in pressure, temperature, and gas composition.

CHAPTER 5

The Components of a Gas Storage Facility

A storage facility usually consists of some of the following components:

Underground reservoir
Injection wells
Withdrawal wells
Injection-withdrawal wells
Observation wells
Gathering system
Compressor facility
Metering facility
Dehydrator
Transmission line to the pipeline

Underground Reservoirs

Underground reservoirs are geological structures that have unique features. There is a porous medium having some degree of permeability. The porosity allows natural gas to be contained within the medium. The permeability allows the gas to move from point to point within the medium. There is almost always an impermeable layer overlying the porous medium. This impermeable layer is usually curved or dome-shaped and prevents the gas contained in the porous medium from rising to the surface of the ground. The curve of the caprock may also prevent the lateral movement of the gas outside the porous medium. In some cases a geologic fault may have produced a vertical shift at one or more sides of the gas sand to provide a lateral seal. The bottom of the porous medium may be sealed by impermeable rock, or it may be sealed by water.

The greater the porosity is of a given medium, the more gas can be contained in a unit volume of the medium. Porosity may vary from zero to some higher number. Porosities of 15 percent are not unheard of. Porosity may vary from point to point within the same gas sand.

Permeability is the ability of gas (or other fluid) to flow through a medium with a given pressure drop. The greater the permeability is, the greater the flow is for a given pressure drop. Because flow in a permeable medium requires pressure drop, the pressure in a reservoir that is being produced will vary within the reservoir. At the well bore the pressure will be the lowest when gas is being withdrawn. Conversely, the pressure will be highest at the well bore when gas is being injected into the reservoir. Permeability can vary throughout the reservoir. In extreme cases permeability can decrease to zero and seal off a portion of the reservoir. This is called a permeability pinch-out. Permeability in underground structures can be as low as zero and as large as several thousand millidarcies. Because of the reservoir's permeability characteristics, the pressure profile in a reservoir may vary widely. When a field is shut in, it may take a very long time for the pressure to stabilize throughout the field.

Underground reservoirs may be classified into two general types: volumetric reservoirs and water drive reservoirs. The volumetric type is sealed on all sides by impermeable rock and behaves like a pressure vessel. This structural seal by geologic characteristics keeps its size and shape constant. The water drive reservoir is sealed on the top and sides by impermeable rock but is sealed on the bottom by water. This type of reservoir may be thought of as a bucket inverted in a body of water. As more gas is introduced into the bucket, water is forced out and the size of the gas bubble inside the bucket increases. The forcing back of the water may occur by two mechanisms. In one mechanism, the gas and a very large amount of water are contained in a large sealed reservoir of constant volume. As gas is injected into the reservoir, the increased pressure causes the water to be compressed. Because the compressibility of water is quite small, a very large volume of water is required in order to achieve an enlargement in the gas bubble. The second mechanism is where the gas simply pushes the water out of the storage reservoir to some other reservoir or some other location.

Figures 5-1 and 5-2 illustrate in a simple manner the way in which the two reservoir types behave. In Figure 5-1, the volume of the reservoir pores that contain the gas is constant. As the quantity of gas in the reservoir is doubled, the absolute pressure doubles. This simplified example assumes constant temperature and ideal gas laws. In Figure 5-2, as the quantity of gas is doubled, the volume of the reservoir pores containing the gas increases due to some of the water being pushed back. This creates a larger storage volume. For this reason the pressure does not rise as rapidly as in the case of the volumetric structure. The increase in pressure is only 75 percent, as opposed to 100 percent in the volumetric case.

As the gas is produced from fields that have a water-drive reservoir, the decreased reservoir pressure allows the water to encroach and slowly fill the reservoir. At depletion, the reservoir may be completely filled with water except for a small gas cap. If the field is converted to a storage facil-

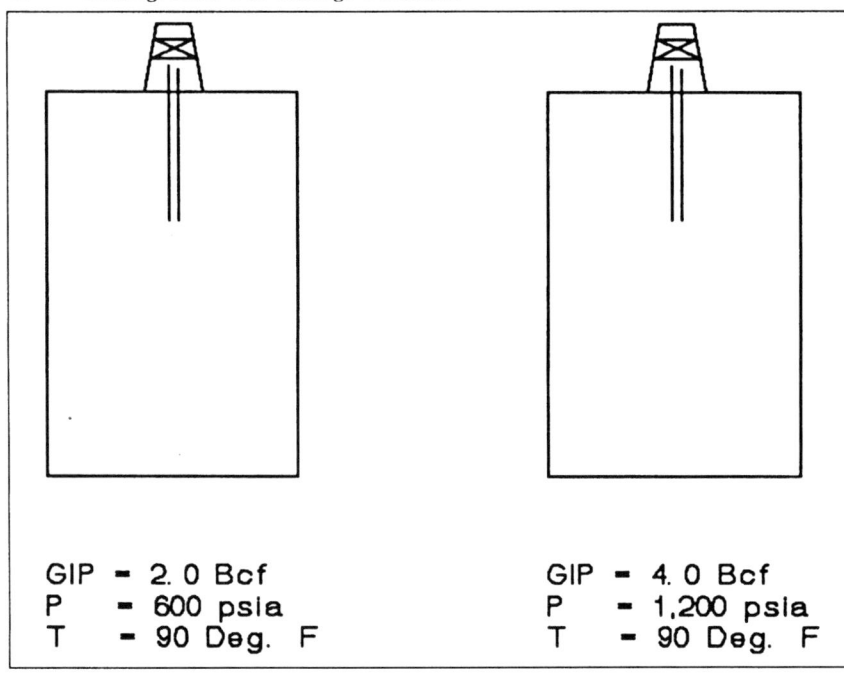

```
GIP = 2.0 Bcf          GIP = 4.0 Bcf
P   = 600 psia         P   = 1,200 psia
T   = 90 Deg. F        T   = 90 Deg. F
```

Figure 5-1. Example of a volumetric reservoir.

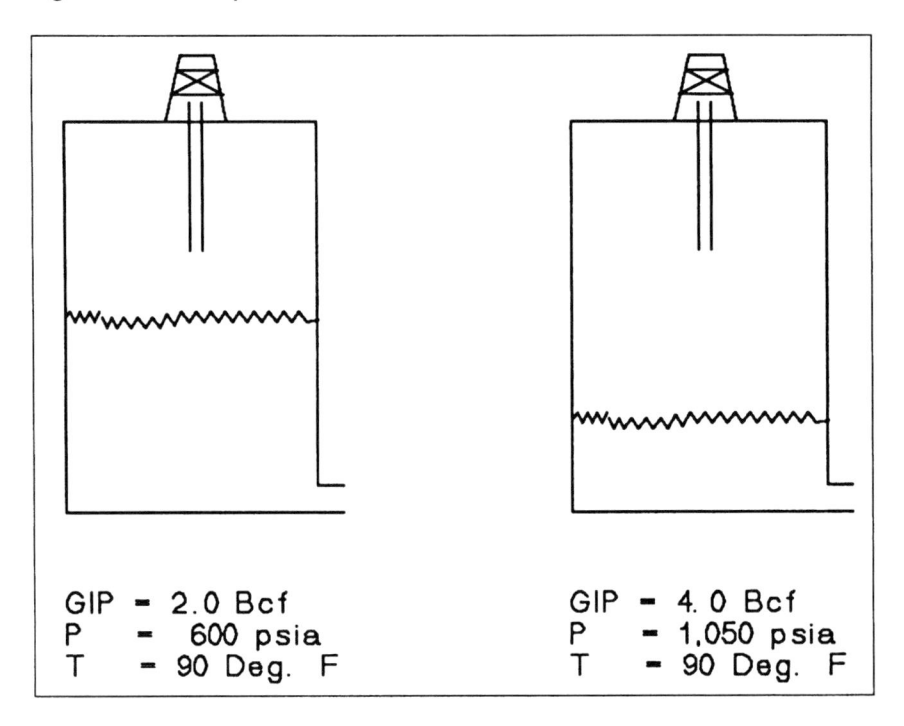

```
GIP = 2.0 Bcf          GIP = 4.0 Bcf
P   = 600 psia         P   = 1,050 psia
T   = 90 Deg. F        T   = 90 Deg. F
```

Figure 5-2. Example of a water-drive reservoir.

ity, new storage wells may be drilled. These new wells will usually be large bore, high capacity wells as opposed to the relatively small bore wells that were used for original production. Therefore these new wells have the capability of injecting gas at a high rate. Care must be used when this is the case. Ideally, when gas is re-injected into a reservoir filled with water, the gas would displace the water and maintain a level, horizontal gas-water interface. If this were the case, gas could be re-injected at a high rate. Unfortunately, the gas has a tendency to override the water and travel along the top surface of the reservoir. This can result in a gas-water interface that is highly sloped and runs down-structure. The more level the cap rock is and the faster the rate of injection, the more pronounced this effect is. Figure 5-3 shows an example of this gas override. There is the possibility that gas may travel along the top of the reservoir and reach a low point water seal. If this occurs, gas may be lost from the storage reservoir and be transmitted to an adjacent reservoir. If the adjacent reservoir is not owned as part of the storage facility, the gas may be lost forever.

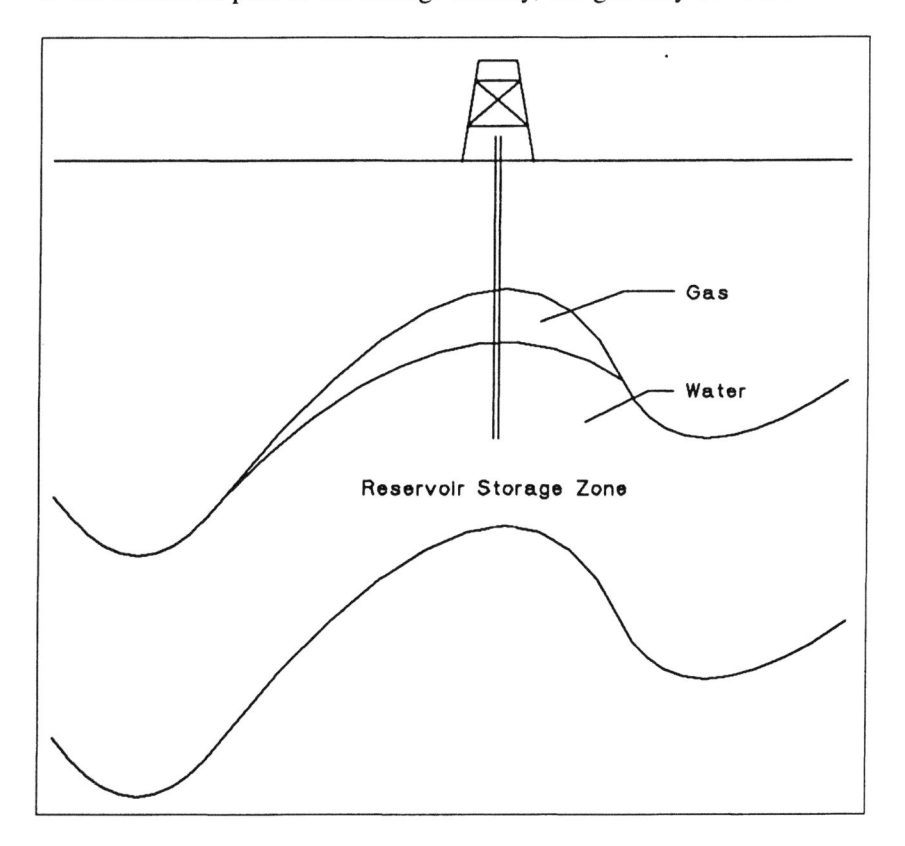

Figure 5-3. Example of gas override in a developing storage reservoir.

Therefore it may be necessary to inject at a relatively low rate during the early stages of reservoir development. Observation wells also may be of help in monitoring the extent and shape of the gas-water interface.

Underground storage reservoirs contain two types of gas: cushion gas and working gas. These two types are functional rather than physical. The cushion gas may be native gas that has never been produced, or it may be injected gas that has been produced at some other location and injected into the reservoir. The cushion native gas may be further classified into recoverable and nonrecoverable gas, depending on whether it is economically feasible to recover the gas. Working gas is almost always injected. For purposes of this text, the only classifications of interest are working gas and cushion gas.

Cushion gas is gas that is necessary to have in the reservoir in order to provide the pressure needed to withdraw the working gas. The cushion gas stays in the reservoir and is not withdrawn. It is a constant quantity. Working gas is withdrawn and injected each year and sometimes several times per year. The amount of working gas in the reservoir can vary from zero to the maximum amount.

Wells

Wells are used to transfer gas into and out of the storage reservoir. Combination injection-withdrawal wells are the most common type because they are the most economical. They are used to either inject or withdraw gas. Due to the individual reservoir characteristics, it may not be feasible or desirable to inject or withdraw in a particular portion of the reservoir. This might be due to a desire to control water influx, the shape of the gas bubble, or other reasons. In those cases there may be one or two wells that are used for only injection or only withdrawal. The injection-withdrawal wells are usually large-bore wells, much larger in diameter than normal producing wells. This increased diameter is to give the increased deliverability needed from storage wells.

Observation wells are used to monitor water migration in a reservoir. There might be a "saddle" or low point in the structure where gas could escape if it reached there. An observation well might be placed on the gas bubble side of the saddle to monitor whether there is gas or water at that point. Observation wells can be of small diameter. Sometimes a dry hole or unsatisfactory injection well will be converted into an observation well.

The following equation may be used to calculate the flow rate for a single well during withdrawal from the reservoir:

$$Q = C \times (P_{SIS}^2 - P_{FS}^2)^n \tag{5-1}$$

Q = withdrawal flow rate
P_{SIS} = shut in surface pressure
P_{FS} = flowing surface pressure
C = coefficient determined by test
n = coefficient determined by test

The characteristic fluid flow equation form suggests that the flow rate is dependent on the terminal pressures and the nature of the well tubing. This is not quite correct for a producing well. The shut-in surface pressure is the bottom hole pressure corrected for the elevation effect due to the depth of the well. This pressure should be taken when the reservoir is stabilized and the pressure is representative of the pressure throughout the reservoir. When the well begins to flow, the bottom hole pressure drops due to the friction caused by gas flow in the reservoir. For this reason the shut-in surface pressure is not indicative of the flowing bottom hole pressure. It is indicative of the reservoir pressure quite a distance from the well.

Similarly, the coefficients C and n do not portray the characteristics of the well tubing alone. They are representative of the flow properties of the well tubing plus the flow properties of the reservoir for some effective distance from the well. Thus two identical wells of equal depth drilled in two different reservoirs would probably have different values for the coefficients C and n.

The conventional method for testing a well to determine the values of the coefficients is to shut in the entire reservoir and allow the pressure to stabilize throughout the reservoir. This stabilization is quite important because the reservoir may have a very non-uniform pressure profile. A long period of time may be needed to achieve pressure stability. Once the pressure is stabilized, a series of tests at four different flow rates is performed. The first flow rate selected should be the lowest of the four flow rates. The flow rate is set and the well is allowed to flow until the flowing surface pressure stabilizes. Ideally this stabilization will occur in about 30 minutes, but in practice it may take several hours. When the flowing surface pressure has become constant, the flowing pressure is recorded and

46 Underground Gas Storage Facilities

the data necessary to calculate the flow rate is recorded. The flow rate is then increased to the next planned point and the procedure is repeated. This procedure is continued until all four tests have been completed. It is usually assumed that the shut-in surface pressure is not affected by the production of gas and remains the same for all four tests.

Traditionally the results have been plotted on a graph similar to Figure 5-4. This is a logarithmic plot of the flow rate versus the difference in the squares of the pressures. The best line was drawn through the points, and

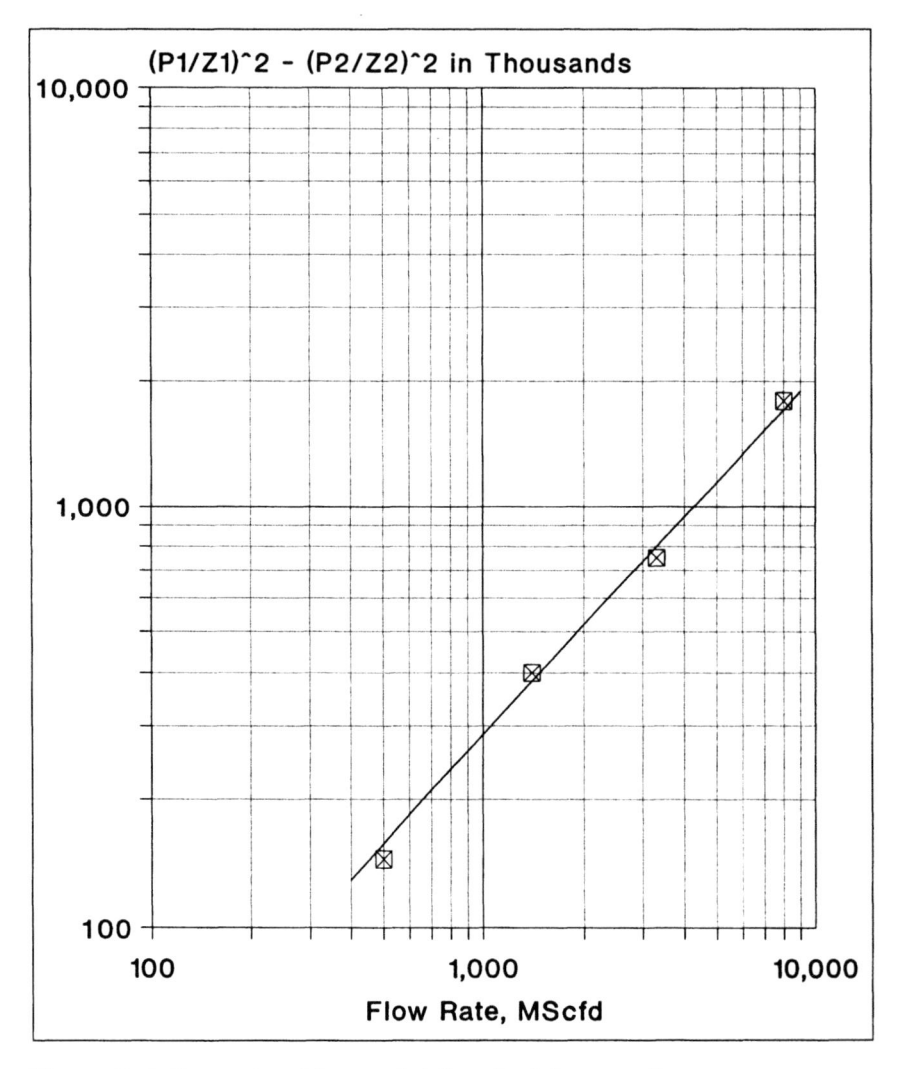

Figure 5-4. Example of four point flow test for a well.

the coefficients C and n were determined from the line. It is not necessary to actually draw the graph. The coefficients can be determined from a least squares correlation of the data. The logarithms of the data must be used in the least squares fit. The coefficients determined in this manner should be valid until something changes in the physical character of the well tubing or the reservoir. As an example, the reservoir sand near the well bore may become contaminated with dirt during injection. This accumulation of dirt may clog the sand and decrease its permeability. This could then affect the value of the coefficients.

In the development of a storage field, it is common to drill a small number (perhaps three or four) of wells in order to predict how many wells will be needed and what the total deliverability of the facility will be. This information is needed for any Federal Energy Regulatory Commission (FERC) application that may be necessary. The usual practice is to make the well tests by injecting gas rather than by withdrawing it. In that case Equation 5-1 is modified slightly:

$$Q = C \, (P_{FS}^2 - P_{SIS}^2)^n \tag{5-2}$$

Q = withdrawal flow rate
P_{FS} = flowing surface pressure
P_{SIS} = shut-in surface pressure
C = coefficient determined by test
n = coefficient determined by test

The data from these test wells can be used to plan a total facility well location and drilling program. The petroleum engineer can use this data along with the estimated sand thickness, porosity, and permeability at each proposed well location to estimate the performance of each proposed well.

There may be considerable uncertainty as to the sand thickness, porosity, and permeability at the proposed future well locations. Because of this, the projections on the performance of these proposed wells may not be completely accurate. There may even be some dry holes which do not produce any gas. Because of these eventualities, it is suggested that the calculated results for the proposed wells be de-rated to provide for a contingency. This may be done by adjusting the C coefficient and leaving the n coefficient alone. The author has successfully used a factor of 0.75 as a multiplier for all of the proposed C coefficients for the wells yet to be drilled.

Depending on the shape of the storage structure in the reservoir, it may be desirable to drill one or more observation wells. These may be slim hole wells, as they will be used primarily to check the level of the gas-water interface. These wells should be located at strategic positions to warn of possible gas loss from the storage sand.

The wellhead facilities should consist of a liquid separator and a pig trap as a minimum. The separator is necessary to remove any produced or entrained water from the reservoir. Unless the reservoir is very dry, the withdrawal gas will usually bring with it some salt water. It is important to remove this salt water before it gets into the gathering system.

Gathering System

The gathering system connects the system of wells with the central point facilities. This gathering system is similar to the gathering system in a producing field, except that the gathering lines may be larger in order to accommodate the larger volumes from large-bore wells. The equation used to define the flow of gas in a pipe is:

$$Q = 0.001368 \times \left[\frac{(P_1^2 - P_2^2) \times D^5}{G \times T \times L \times f} \right]^{0.5} \tag{5-3}$$

Q = flow rate in MMScf per day
P_1 = upstream pressure, psia
P_2 = downstream pressure, psia
D = pipe inside diameter, inches
T = gas temperature, °Rankine
L = pipeline length, miles
G = gas specific gravity
f = Fanning friction factor

This equation assumes that there are no significant changes in the elevation of the pipeline and that the standard conditions for the gas are 60°F and 14.73 psia. There are several simplified correlations that have been developed for the Fanning friction factor. While most of these work well over a limited range, they can become quite inaccurate over a wide range of conditions. It is suggested that the Colebrook-White correlation or the Moody diagram be used to obtain the friction factor.

The most common arrangement for a gathering system is the tree configuration where wells are connected to pipes, which are in turn connected to successively larger pipes. A meter is installed at the wellhead. There have been some instances where individual lines have been used to connect each well to the central point. The individual well meters are then located on each line near or at the central point. Although this is a more convenient arrangement for checking the flow to and from the individual wells, it is more costly and is not widely used.

It is usually desirable to install a meter to measure the flow rate of gas to and from an individual well. This is often necessary in the management of the reservoir. In cases where the flow rate from or to an individual well is quite high, it may be necessary to install a flow limiting device to prevent damage to the sand face near the well.

It is strongly recommended that good quality liquid separators be installed at the wellhead unless the reservoir is known to be nonproductive of water. Water can be produced periodically from a reservoir during withdrawals. This water can contain substantial quantities of dissolved substances such as salt. If allowed to get into the pipeline, the liquid can coat the walls of the pipe. The flow of gas will then dry up the water over time, leaving a residue of very fine grained salt. Due to the fine grain size, this salt can often go through the central point separator and into the dehydrator, contaminating the glycol. If the salt concentration in the glycol gets high enough, the glycol will have to be replaced, resulting in additional expense and down time.

Compressors

The compressor facility is usually located at some central point near the wells and may be used to compress the gas for injection, for withdrawal, or both. The compressors are generally used for injection because the reservoir operating pressure is usually higher than the transmission system pressures. Since the compressors are available, they are often used for withdrawal also in order to increase deliverability. There are cases where a shallow, low-pressure field is used for gas storage and the injection is done at pipeline pressure and compressors are used to withdraw the gas.

Compressors used for injection service are almost always reciprocating-type compressors. At the beginning of the injection season the discharge pressure of the compressor will be relatively low. Near the end of the season the discharge pressure will be relatively high. A centrifugal compressor normally cannot meet this range of requirements. Centrifugal compressors have been used for withdrawal only service, where the range of suction pressures is not great due to the reservoir being low pressure and having an active water drive.

The basic, generalized equation for reciprocating compressor theoretical horsepower is:

$$HP = Q \times T_1 \times Z_1 \times \left[\frac{k}{k-1}\right] \times \left[\left(\frac{P_2}{P_1}\right)^{\frac{k-1}{k}} - 1\right] \qquad (5\text{-}4)$$

HP = theoretical horsepower
Q = flow rate in MMScf per day
P_1 = suction pressure in psia
P_2 = discharge pressure in psia
T_1 = suction temperature in degrees Rankine
Z_1 = compressibility at suction conditions
k = ratio of specific heats of the gas

The above equation is modified in various ways in order to convert the theoretical horsepower to actual horsepower. Sometimes a number of adjustments are made to correct for various conditions.

The compressors are usually sized for the injection service, as this is usually the more demanding case. In working with reciprocating compressors it should be kept in mind that the horsepower rating on any compressor produced by any manufacturer is a rating for the compressor when it is new, clean, and under essentially ideal conditions. After a few weeks of operation the compressors are no longer in ideal condition. This is true even of well-maintained units. Further, the gas properties may vary from the design basis, causing a decrease in capacity. It is recommended that a de-rating factor be used in sizing the compressors and that the same factor be used in predicting performance in service once the compressor units have been chosen. While individual experience may vary, a de-rating factor of 0.95 is reasonable. When the rated horsepower is multiplied by this factor, the result is a power that is more realistic.

During the injection season the ideal scenario is to inject at a constant rate just the amount of gas necessary to fill the reservoir by the end of the season. It should be kept in mind, however, that operating and market conditions may make it desirable to inject at a higher or lower rate than the average. The storage facility is a convenient place to dump some gas when the pipelines are full. Similarly it can be a source of supply, even in the summer season. These situations should be considered by each individual company during the sizing of compressors to determine if it is desirable to provide some extra horsepower. Since the compressor system will be operating for a significant portion of the year, a spare compressor unit may be prudent.

When the injection horsepower is determined, it is usually economical and desirable to use it for withdrawal service also. This dual service can impose some challenges to the compressor engineer in the selection of cylinder sizes and unloading sequences. The result is usually something that may not be ideal, but is workable.

Central Point Metering

Accurate metering at the central point is essential for good inventory control. Due to the widely different characteristics of the injection and withdrawal flow streams, it is often impractical to measure both streams with a single meter facility. Separate metering facilities for injection and withdrawal are then required. It should be kept in mind during the design of the injection meters that there may be times when it will be desirable to inject gas at relatively high rates. If the injection meters are designed to allow only the average rate of injection, the higher rates may not be possible.

One of the biggest potential sources of metering error is pulsations from reciprocating compressors. A case has been documented where a single production well with a compressor that had no pulsation control equipment was producing into an orifice meter and the result was an error of 56 percent in the meter. Adequate, well-designed pulsation control equipment is essential in order to produce satisfactory metering. The Pipeline and Compressor Research Council (PCRC) of the Southern Gas Association (SGA) has engaged in a large amount of research on this problem and has developed methods for the control of compressor pulsations. The result of this research is embodied in their design facility at Southwest Research Institute in San Antonio.

Central Point Separators

Central point separators should be used for both the injection and withdrawal streams. The injection separator prevents any dust and particle matter brought in from the pipeline from contaminating and clogging the sand face at the well bores. The withdrawal separator keeps any sand from the reservoir from entering the pipeline. Both separators protect the compressors.

The withdrawal separators should be designed to handle liquid as well as particle matter. While it is highly undesirable for liquid to enter these separators, it sometimes happens during withdrawal. Any liquid entering the units during withdrawal will probably be salt water, which will wet the sock filters. When these socks dry out, the residue salt crystals are very fine and can pass on to the dehydrator. The accumulation of salt in the dehydrator glycol can cause the salt to deposit out on the fire tubes in the glycol regenerator, which can cause tube failure. In cases where salt water enters the separator on a habitual basis, it may be necessary to add a high-pressure pump to pump fresh water to spray over the sock material and remove the salt water before it crystallizes. It is much better to contain any salt water at the wellhead before it enters the gathering system.

The gas in a storage reservoir is at moderate temperatures. Typical year-round temperatures may be 80 to 110°F. When this gas is in contact with the water in the reservoir, the gas becomes saturated with water vapor corresponding to that particular temperature and pressure. When the gas is withdrawn from storage, it is usually in the winter period when the ground temperature is perhaps 45°F. This colder ground temperature cools the gas and causes liquid water to condense in the gathering lines. If the gathering system is in level ground, this condensed water is carried along with the gas as it is formed and is taken out in the central point separator. If the terrain is hilly, however, the liquid water will accumulate in the valleys between hills. This accumulation will continue until the pressure drop across the valley is great enough to move the water. When this occurs, the water moves as a slug. When this occurs, the central point separator will not be adequate to contain the large slug, and liquid water will pass on through to the dehydrator. Under these conditions a slug catcher may be required to take out the large slugs ahead of the central point separator.

Central Point Dehydrators

A storage reservoir almost always contains some water and may have an active water drive. When dry gas from the pipeline is injected into the storage reservoir, liquid water from the formation will evaporate into the injected gas. The gas will then have too much water to be pipeline quality gas. The gas must be dehydrated on the withdrawal cycle.

Practically all storage facility central point dehydrators are glycol units. Glycol dehydrators are economical, they have adequate performance, and they operate well unless hit by a slug of liquid water. Although it is very unlikely that a slug of liquid water will ever enter a dehydrator in a well-designed facility, it can effectively shut the dehydrator down if it does. The water dilutes the glycol to the point where its drying capability is lost, and wet gas exits the unit. The remedy is to operate the regenerator until the glycol purity has been restored, which can take 10 to 12 hours. During this time no gas can be withdrawn from the reservoir because it will cause the pipeline to freeze in cold weather.

Periodic checks should be made of dissolved matter in the glycol, particularly sodium chlorides, and a log should be kept. If the salt content of the glycol continually increases, efforts should be made to locate and contain the source. The other main operating variable of note is the regenerator temperature. This should be kept at the manufacturer's recommendation, which is currently about 400°F for most models. When the regenerator temperature falls below the recommendation, improper dehydration and non-specification gas can result.

Transmission Line

The transmission line connects the underground storage facility central point to the pipeline system. This transmission line may be less than a mile long, or it may be several miles in length. It is essentially identical to all the other transmission lines in the system, and is designed in a similar manner.

CHAPTER 6

Characteristics of Underground Storage

Chapter 5 discussed the various components that make up a storage facility. Each of these components has its own characteristics. When the components are melded together into a unit, the unit takes on some of the individual characteristics of the components and all of the limitations. The unit then has a unique set of characteristics. Because of these unique characteristics, some criteria are needed by which to rate storage facilities. Currently two criteria are used. One criterion is deliverability. This is the maximum amount of gas that can be withdrawn on a daily basis. The other is working gas capacity. This book suggests a third criterion that combines the first two.

Storage Characteristics

Deliverability. In order to discuss deliverability, the reservoir and the wells will be considered as a unit. When any reservoir first starts producing, the pressure in the reservoir is at the maximum. As the reservoir continues to produce, the continued production causes a decrease in the reservoir pressure. This in turn causes a decrease in the deliverability. When all of the recoverable gas has been produced from the unit, the deliverability has been reduced to near zero. This is illustrated by Figure 6-1. The x-axis is the percent of the total gas in place that has been withdrawn. The curve comes to near zero at 85 percent of the total. This is because some of the gas in place is regarded as nonrecoverable. A general rule of thumb is that about 15 percent of the total gas in place in a new field is nonrecoverable. Some liberties have been taken with Figure 6-1. First, this figure has been drawn for a reservoir that has been drilled with wells that are suitable for storage use. Because of this, the production rates are higher than are normally encountered in a producing field. Second, actual operation of a producing field such as this would require compression at some point, result-

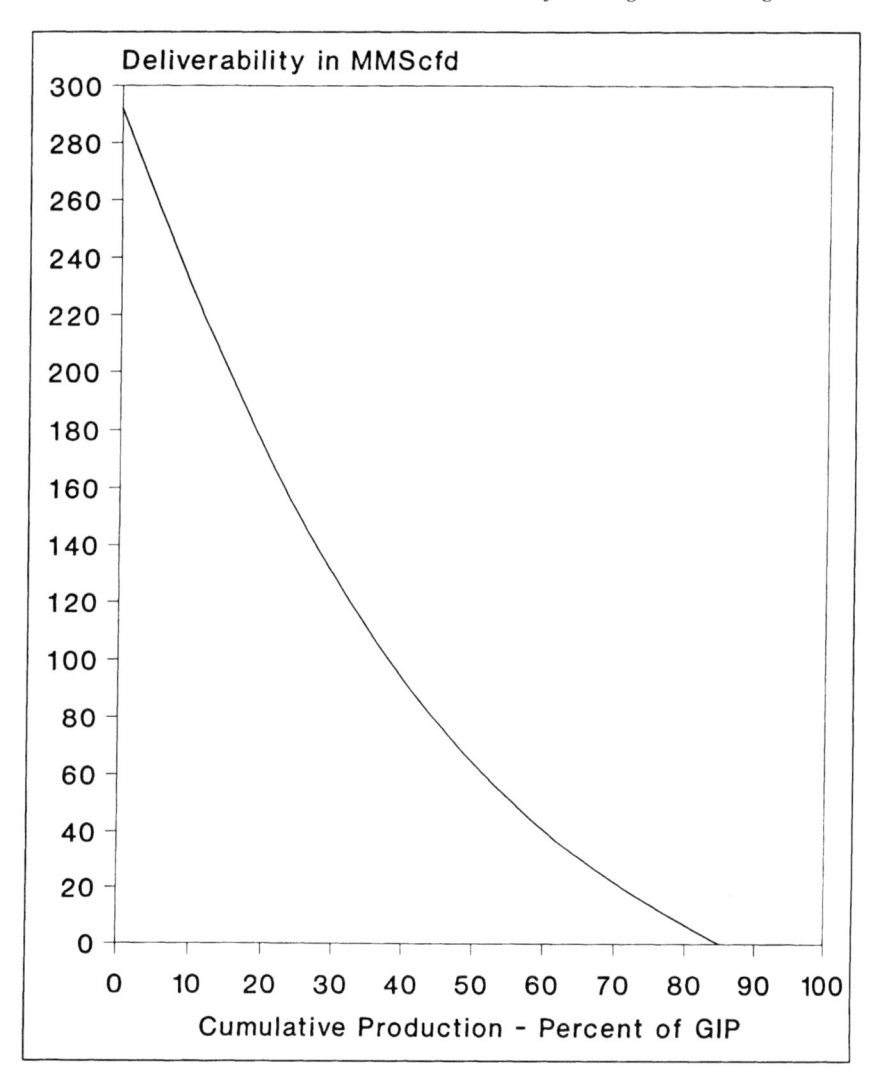

Figure 6-1. Deliverability pattern for a typical reservoir.

ing in a step increase in the deliverability. For simplicity the compression step has been ignored.

The relationship between cumulative production and deliverability is not a straight line function, although the deviation is not severe. The reason for this lack of linearity is due to the nonlinear nature of the well-flow equation. The reservoir pressure is squared in this equation. At the higher pressures when the reservoir is full, the square of the reservoir pressure is a very large number. Any change in the pressure causes a large change in

56 Underground Gas Storage Facilities

the deliverability. At the lower pressures the reverse is true, and a change in pressure causes a relatively small change in the deliverability.

When the working gas volume has been determined for the unit, the same type of relationship holds true. Figure 6-2 illustrates this. In this case the deliverability shown is for the total storage facility, not just the reservoir. When the storage field is full, the pressure is high and the deliverability is 292 MMScf per day. When all of the working gas has been withdrawn, the deliverability is 125 MMScf per day. Thus the deliverability of

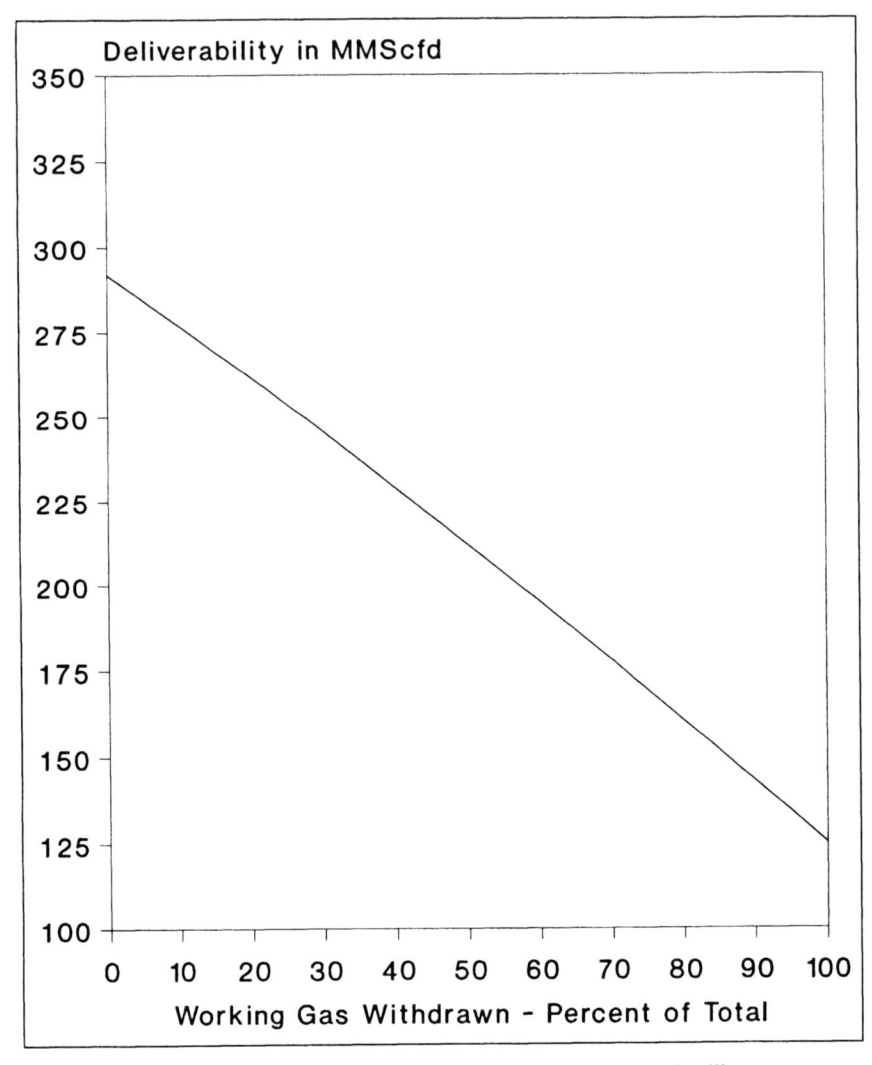

Figure 6-2. Deliverability pattern for a typical storage facility.

the facility varies greatly during the course of the withdrawal season. This raises a question of how to rate the deliverability of the facility.

The unit could be rated at 292 MMScf per day to reflect the maximum at the start of the winter season. The fallacy of this reasoning is that the facility would have an actual deliverability capacity of 292 only on the first day of withdrawal. On the second day, and any following days, the deliverability would be less than the rating. This could produce some anomalies with someone unfamiliar with the rating system and could possibly result in some overconfidence in the storage facility capability. The system could also be rated at the average deliverability for the withdrawal period, which would be 209 MMScf per day. This practice is certainly better than the first option, but it still leaves the deliverability capacity overstated for the last half of the season. It is suggested that the rating be the deliverability for the last Mcf of working gas in the facility. In the example shown, this rating would be 125 MMScf per day. This is a conservative rating that could be met or exceeded at any time during the winter season. In this book all references to deliverability ratings will be based on this method of rating.

Working Gas Capacity. The total gas in a storage field may be divided into working gas and cushion gas. Other names have been used for these two quantities. The working gas is sometimes called top gas, and the cushion gas is sometimes called bottom gas or base gas. For this book they will be called exclusively working gas and cushion gas. The working gas capacity of a specific storage facility is inversely proportional to the cushion gas volume, which is shown in Figure 6-3. This figure assumes that the total capacity of the reservoir is used. If only a portion of the total capacity is used for the storage facility, the same relationship would hold true for the portion of the reservoir that is used. The final result, however, is that the total gas used for the storage facility must be either cushion gas or working gas. The more cushion gas there is in the reservoir, the smaller the working gas volume will be.

The quantity of cushion also affects the deliverability rating. If the total capacity of the reservoir is used for either cushion gas or working gas, the quantity of cushion gas does not affect the deliverability of the facility as long as there is working gas in the reservoir. The quantity of cushion gas does, however, affect the deliverability of the last cubic foot of working gas that is withdrawn. Since this is the rating of the facility, the quantity of cushion gas affects the deliverability rating of the facility. Figure 6-4 shows this. If there were no cushion gas, the facility would produce until

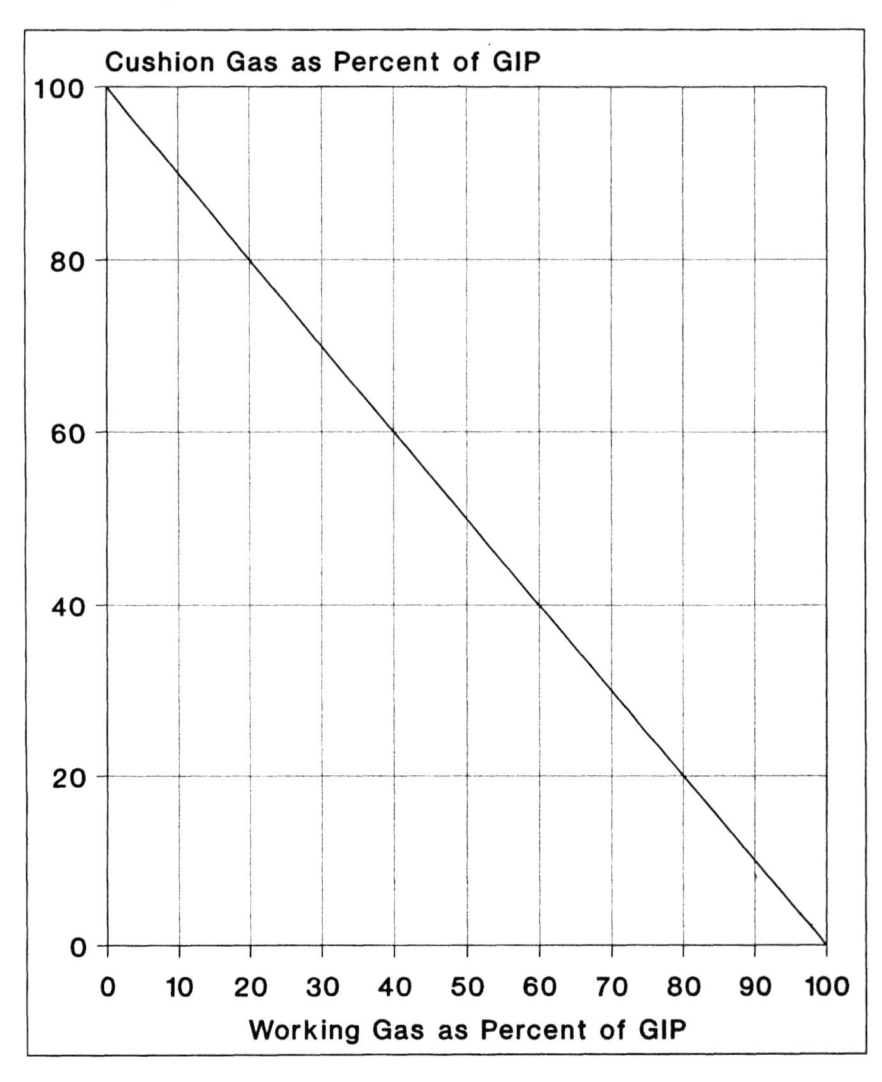

Figure 6-3. Relationship between working gas and cushion gas.

the total recoverable gas in place was withdrawn. At this point the deliverability would be zero, and the rating of the facility would be zero. This example shows one extreme. At the other extreme, assume that the cushion gas was 100 percent of the total gas in place. This would mean that there was no working gas and the deliverability rating would be 292 MMScf per day, but there would be no working gas to withdraw. These two extreme examples serve to show how one quantity affects another. The more reasonable cases occur between these two extremes. While

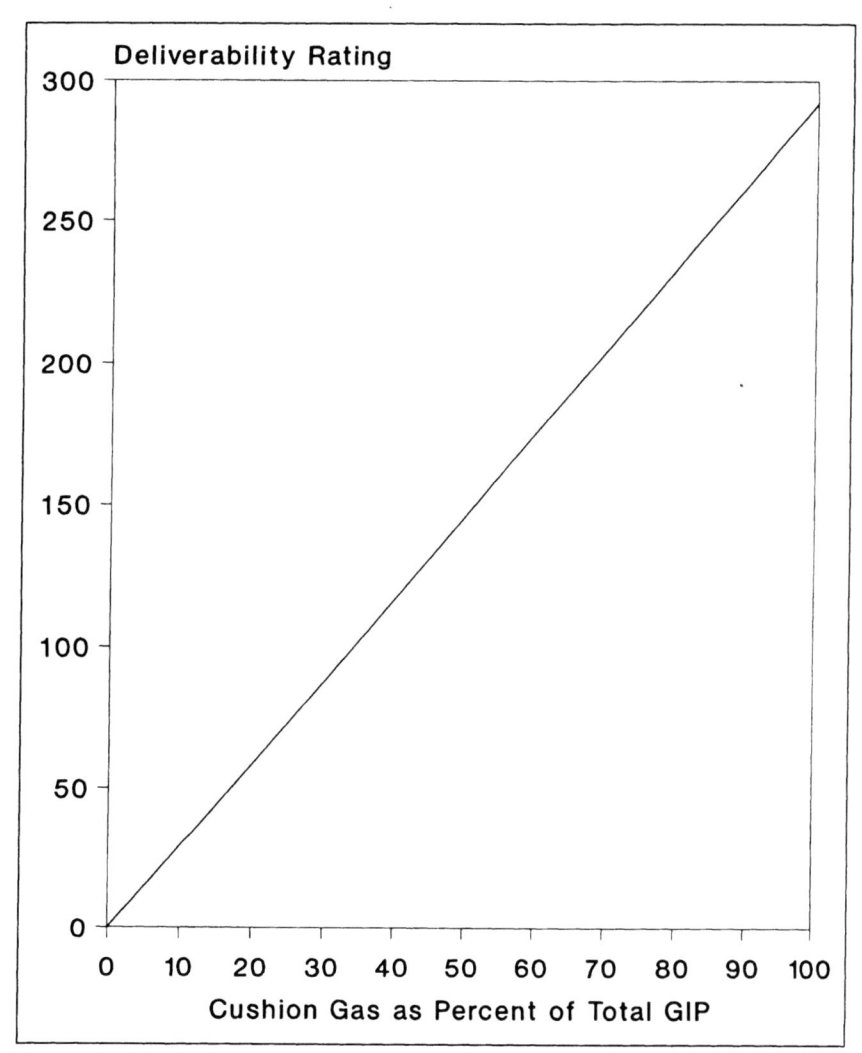

Figure 6-4. Effect of cushion gas on deliverability rating of a storage unit.

there are no requirements or prohibitions here, the range of cushion gas probably varies between 40 percent and 70 percent of the total reservoir capacity for most storage fields.

Types of Storage

Storage facilities are generally divided into two classes: base load storage and peaking storage. The in-between type should probably be another classification. Base load storage is a storage that is brought on early in the winter season as soon as storage is needed. This type typically stays on for several days at a time. It is characterized by relatively large working gas capacity and relatively lower deliverability. The base load storage can be thought of as first-on, last-off storage.

A peaking storage facility, by contrast, usually has a somewhat limited working gas capacity and a rather high deliverability. This type of storage is the "last resort" storage. It is usually the last storage brought on and the first storage to be taken off the line. Because of its limited working gas capacity it must be used judiciously. Peaking storage is used on the coldest winter days to fill in the peaks on the load curve that are left empty by the normal supply and the other storages.

Although these are the typical storage types and usages, there can be considerable variety in actual usage. For example, a peaking facility that is highly automated and can be brought on and off the line easily and conveniently may be used early in the season on a daily basis simply because it is convenient for the gas dispatcher to use. The facility can be used to fill in gaps of a few hours and then be switched to injection service to refill the reservoir. Similarly, this facility would be a convenient place to dump gas when a market suddenly disappears. By contrast, a storage that is jointly owned with another corporation may not be as convenient to bring into injection or withdrawal service as one that is totally owned. A storage service that is rented from another company may require several hours' or several days' notice before using. These storages would tend to be not used until it could be assured that they would be on line for a relatively long period of time. Convenience of use by the gas dispatcher becomes the governing factor for use in many cases.

When base load and peaking service is discussed, a criterion is needed to determine which is which. This book suggests that a new measurement be introduced to assist in this. Earlier it was suggested that the deliverability rating of a facility be the deliverability for the last Mcf of working gas in the reservoir. Following this convention, it is suggested that the time to empty the working gas from the reservoir at the rated deliverability of the facility be used as a criterion for typing the facility. A storage reservoir that had a working gas capacity of 8 Bcf and a deliverability rating of 100 MMcf per day would require 80 days to empty at the rated deliverability. From examination of several storage facilities in use it can be determined that base load facilities usually have characteristics that require about 70 to 120 days to empty using this definition. Similarly, peaking units require 5 to 25 days to empty the working gas. Using this background it is suggested that the following classification be used:

- Peaking. Facilities that require less than 30 days to empty the working gas at the rated deliverability.
- Mid-range. Facilities that require 31 to 80 days to empty the working gas at the rated deliverability.
- Base load. Facilities that require more than 80 days to empty the working gas at the rated deliverability.

The exact numbers that define these ranges may be open to discussion. Indeed, the defining numbers are not of overwhelming importance. The key point is to have some ranges with which to work.

One fallout of the classification criteria listed above is the maximum time required to empty a reservoir at its rated deliverability. Most storage facilities have emptying rates of 120 days or less. When a proposed facility has an emptying rate of 150 or 175 days or more, considerable thought should be given as to whether this facility will serve the needs for which it is intended.

In the past the conventional method of operating a storage reservoir was to fill it up during the summer season and to empty it in the winter season. In recent years it has become customary to inject into the reservoir during the winter season when the weather warms up and excess gas is available. This practice is particularly prevalent on peaking facilities. These units usually have smaller working gas capacities. The recharging of these storage units during the winter season gives them a higher effective working gas capacity. In some cases it can even be advantageous to use gas from a base load storage to inject into a peaking facility.

Tailoring of Storage Facilities

Storage reservoirs have some inherent characteristics that cannot be circumvented. In spite of this, however, the various components of a storage facility can be modified to tailor the facility to some degree to meet the needs of a given situation. A reservoir may not be permeable enough to give it the deliverability normally desired for a peaking facility if it is developed in the conventional manner. Some measures can be taken that can help alleviate this limitation. For example, more wells may be drilled to increase deliverability; a larger percentage of the reservoir may be used for cushion gas; or a lower wellhead pressure may be used, requiring more compressor horsepower. These measures may not be the optimum for the normal development of the reservoir, but they may be the optimum for obtaining the particular facility desired.

62 Underground Gas Storage Facilities

Figure 6-5 shows graphically how the wellhead pressure and the amount of cushion gas may be used to tailor a storage facility to be either a peaking facility or a base load facility. This graph is for a reservoir that has a total gas-in-place capacity of 22.1 billion standard cubic feet. Four preliminary wells have been drilled and tested. The information from these tests has been used to lay out a drilling pattern for an additional 12 wells. The properties of the future 12 wells have been estimated by the petroleum engineers by conventional methods, and the well coefficients

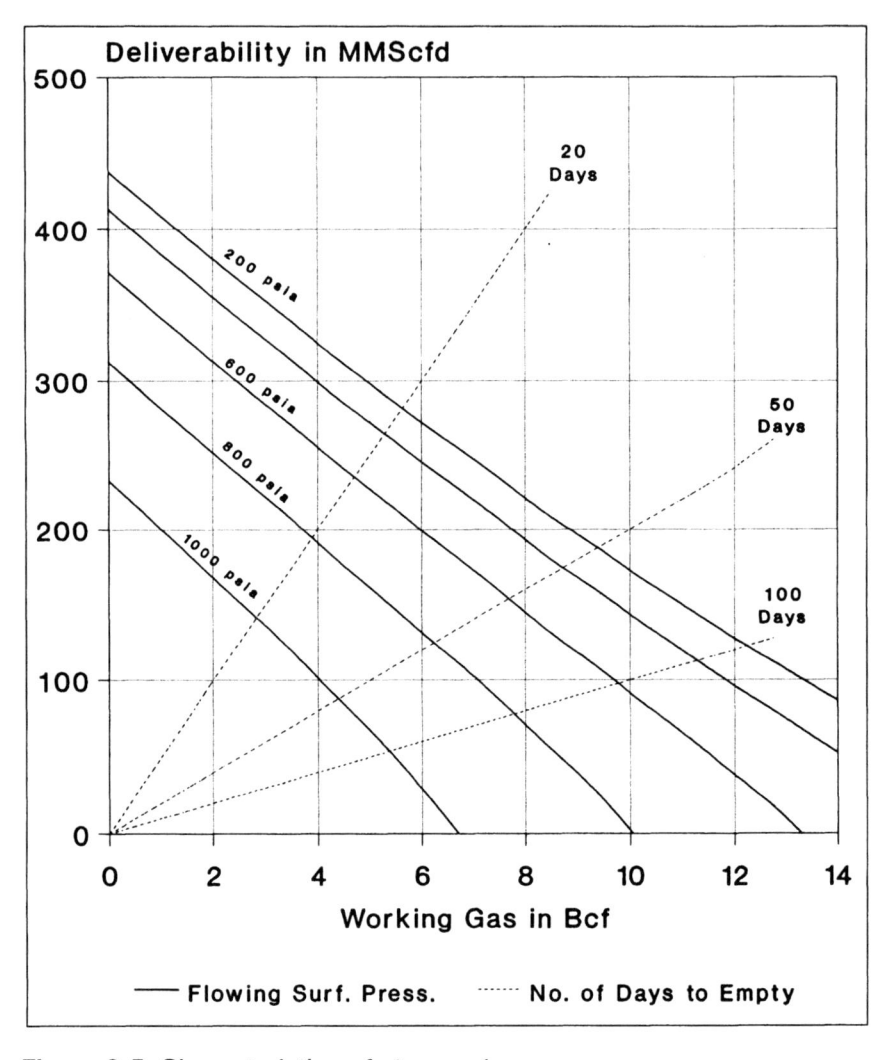

Figure 6-5. Characteristics of storage 1.

Characteristics of Underground Storage

have been de-rated for uncertainty and contingency as suggested in Chapter 5. When the reservoir is full of gas the shut-in surface pressure is 1365 psig. A pressure decline curve was calculated to show the pressure as the gas in the reservoir is withdrawn. All this information was used to calculate the information in Figure 6-5. Figure 6-5 plots the working gas amount versus the deliverability, with flowing wellhead surface pressure as the parameter. Also plotted on the figure are lines of constant time for emptying the reservoir at the rated deliverability. If it were desired to make a base load storage out of this facility, some choices are available. A base load unit with a working gas capacity of 12.1 Bcf could be developed with a flowing wellhead surface pressure of 200 psia and a rated deliverability of about 121 MMScf per day. This would require a certain amount of compressor horsepower and would require 100 days to empty at the rated deliverability.

A second choice would be a working gas capacity of about 9.8 Bcf with a flowing wellhead pressure of 600 psia. This would give a rated deliverability of 98 MMScf per day and would require less compressor horsepower. The time to empty would still be 100 days. A third choice would be a working gas capacity of 7.8 Bcf with a flowing wellhead pressure of 800 psia. This would give a rated deliverability of 78 MMScf per day and would require even less horsepower for withdrawal. Thus the development of the components into a base load storage could be a decision based on the economics and the particular storage needs.

If it were desired to develop the reservoir into a peaking facility, a similar set of choices is available. Using the 20 days to empty line as a guide, some alternates could be considered. A peaking facility could be developed having 5.6 Bcf of working gas with a rated deliverability of 280 MMScf per day. The flowing wellhead pressure would be 200 psia. Another choice would be a facility with 3.9 Bcf of working gas, a rated deliverability of 195 MMScf per day, and a flowing wellhead pressure of 800 psia. Other choices could be explored. Although this reservoir is probably best suited as a base load or mid-range storage facility, the figure shows that a considerable amount of tailoring can be done.

Figure 6-6 shows a reservoir that has a total gas in place of 13 Bcf and a maximum pressure of 1365 psia. A similar set of choices exists here as on Figure 6-5. An examination of Figure 6-6 shows that the reservoir is probably best suited for a peaking type facility. Nevertheless a very suitable base load facility could be designed from it. An example would be a working gas capacity of 9.0 Bcf, a deliverability rating of 90 MMScf per day, a flowing surface pressure of 200 psia, and a time to empty of 100 days.

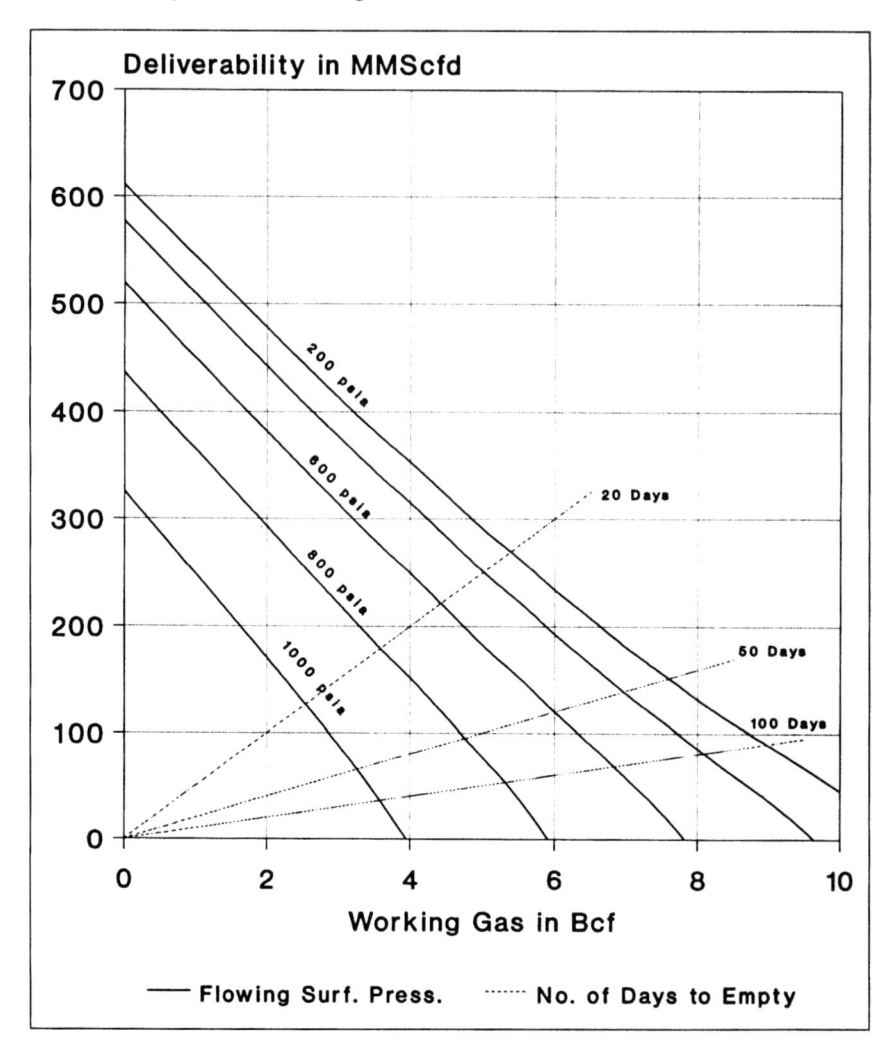

Figure 6-6. Characteristics of storage 3.

These examples show that considerable leeway can be obtained in the final result by using the various parameters to tailor a reservoir to a particular need. The choices are governed sometimes by the needs at hand and sometimes by economics. The key point is to be aware that choices are available.

Sizing of Surface Facilities

Figure 6-2 shows the deliverability pattern for a typical storage facility. The facility referred to by this particular figure has a deliverability of 292

MMScf per day at the beginning of the withdrawal season. When all the working gas has been withdrawn, the deliverability has decreased to 125 MMScf per day. The question arises as to how to design the surface facilities to accommodate this variation. If the dehydrators, separators, and other equipment are designed for the rated deliverability of 125 MMScf per day, a large portion of the facility's capacity cannot be used. On the other hand, if the surface equipment is designed for the maximum deliverability of 292 MMScf per day, the facility will have this capacity for only one day. After this first day, the capacity will be less. This would be an expensive choice and would waste auxiliary equipment capacity. Clearly something in between is indicated.

In Chapters 1 and 2 the severity and timing of individual daily temperatures was discussed. A study of some actual weather station data at selected locations showed that for these particular locations, the coldest day of the year occurred on January 12 on a statistical basis. Keep in mind that this date of January 12 is based on statistical averages over a long period of time, and that in any given year the coldest day of the year may deviate considerably from this date. Nevertheless it is logical that, on an average, the coldest day of the year would occur about halfway through the winter period. It would also seem logical that the capability of a storage facility might be best utilized if it were geared to this particular peak day capacity.

If it is assumed that the use of storage working gas will be in a relatively symmetrical pattern throughout the winter period, then about half of the working gas will have been withdrawn halfway through the winter period, which would also coincide with the peak day on an average basis. Therefore it is proposed that the surface facilities related to the withdrawal cycle be sized for the deliverability of the facility at the midpoint of the withdrawal cycle. For the facility represented by Figure 6-2, this capacity would be 209 MMScf per day. This would include the wellhead equipment, the gathering system, the central point separators, the dehydration equipment, the metering equipment, and any auxiliaries. This would mean that the facility would be limited to a capacity of 209 MMScf per day during the first half of the withdrawal cycle even though the reservoir would be capable of more. It is questionable whether there would be a need for more capacity than this during this period. At any rate this limitation represents an economic compromise. The withdrawal capacity of the facility throughout the season is represented by Figure 6-7.

The sizing of the injection facilities should also be addressed. Assume that a storage reservoir has a working gas capacity of 12 Bcf, and that it

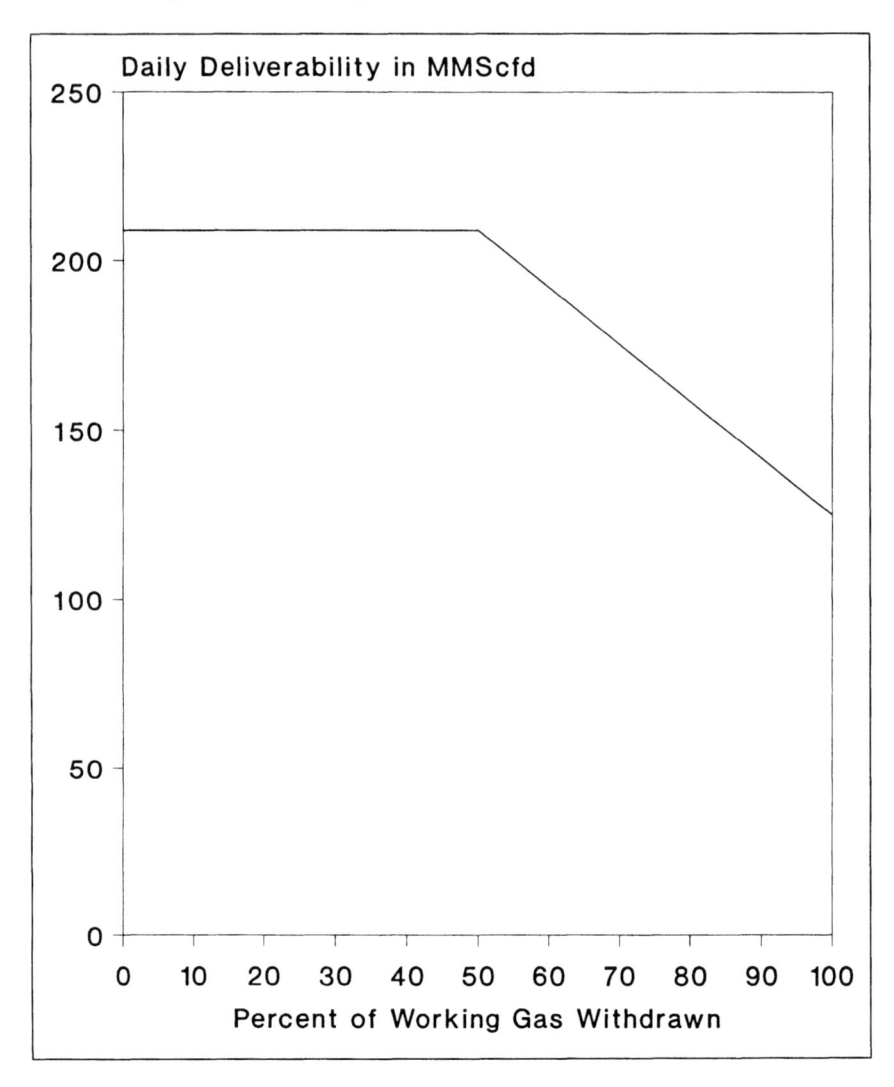

Figure 6-7. Deliverability of storage facility with proposed surface capacity.

is planned to withdraw all of this working gas each winter season and to reinject it each summer season. The withdrawal season lasts for about five months. Assume a shut-in period of 30 days each year for inventory monitoring. This leaves six months or 180 days for injection. This would require that 67 MMScf per day be injected each of the 180 days in order to replenish the 12 Bcf of working gas. Often this is the rationale that is used to design the injection facilities.

Some other considerations should be examined, however. The compressors are probably used for both injection and withdrawal, so that they see service all year. Although a spare compressor and a preventive maintenance program should be in place, there will be cases of unanticipated compressor breakdowns that will restrict the normally planned injection rates. In addition there can be times where it will be advantageous to inject gas at higher than scheduled rates. This can occur when there is excess gas on the system and it is simpler to inject the extra gas into storage rather than cutting it off in the field.

For these reasons it may be desirable to size the injection facilities for higher rates than those shown by the simple average of working gas capacity and number of injection days. The equipment affected by this will be the separation equipment and the metering equipment. The compression equipment can probably handle the extra capacity by using the spare compressor, although this will be an operating and economic decision.

CHAPTER 7

Optimization of Underground Storage Facilities

The investment in an underground storage facility is large. The cost is in the tens of millions of dollars and sometimes in the hundreds of millions of dollars. It is prudent to attempt to manage these investment costs consistent with good judgement to achieve the lowest annual cost over a long period of time. As with many other types of projects, the minimum investment does not always translate into minimum annual costs.

The amount of flexibility to achieve optimum economic conditions will vary with each case. There will be times when the requirements of the storage needs and the limitations of the reservoir chosen will be so confining that there will be little room for economic maneuvering. Nevertheless it is well to examine some of the conditions that govern costs.

Table 7-1 lists some of the factors that affect storage facility characteristics and thereby affect the economics. Once a reservoir is chosen, the reservoir capacity and the maximum reservoir pressure are fixed. There is always the option of making use of only a portion of either of these. This option is rarely exercised, however. The economics of using only a portion of a reservoir are usually not attractive. The exception would be when a moderate storage is needed and only a large reservoir is available, and the need for storage is so great and the alternatives so undesirable that the normally poor economics become viable.

The variable factors are the number of wells, the flowing wellhead pressure, and the working-gas to cushion-gas ratio. Other factors are variable to some degree. These include the size of the gathering system piping and the size of the transmission line connecting the storage facility to the rest of the transmission system. Good design principles, however, usually dictate to a large extent what the sizing on these components will be. Sometimes a reduction in cost can be gained by trading off compressor horsepower with either gathering system cost or transmission line cost.

Table 7-1
Factors Affecting Storage Characteristics

Fixed Factors

    Reservoir Capacity

    Maximum Reservoir Pressure

Variable Factors

    Number of Wells

    Wellhead Pressure

    Working Gas to Cushion Gas Ratio

Storage Wells

With the exception of cushion gas, the largest single cost item in a storage facility is usually the compressor installation. This provides the pressure rise necessary to counteract the various pressure drops in the system. The largest single source of pressure drop is the storage wells. It would seem that there would be a trade-off between number of wells and the amount of compression installed. Unfortunately this trade-off is not always utilized. The compression equipment is often sized for the injection service. Although the compressors may be used for withdrawal of the gas from the reservoir, the amount of compression needed for injection is taken as a given and used to provide the amount of extra withdrawal deliverability that can be achieved. The number of wells, on the other hand, is usually governed by the withdrawal service. For this reason any investment trade-off that can be achieved between the number of wells and the amount of compression required is often not pursued. This chapter will look at the economic aspects of compression on both the injection and withdrawal cycles.

Determining the number and size of wells in a storage facility has traditionally been the province of the petroleum engineer and the production engineer. Certainly there are aspects of the wells that are best handled by these disciplines. Determining the cost factors in the drilling and sizing of wells is so specialized that the production engineer is best equipped to handle this area based on previous experience in producing wells. Similarly, the petroleum engineer is in the best position, by virtue of education, to determine what the effective radius of drainage will be for a particular existing well or for a proposed well. The pipeline engineer would be out of his or her natural environment in these specialized areas. The pipeline engineer should, however, be cognizant of these matters and how they affect the overall economics.

As an example, consider the number of wells. Assume that the petroleum engineer has determined how many wells are necessary to adequately drain the reservoir for withdrawal purposes. The question may then arise, "What is the effect of adding one more well?" Normally the question of adding another well is addressed from the viewpoint of increased flow. The petroleum engineer has already considered this aspect. Suppose, instead, that the additional well is considered in light of maintaining the same flow rate but decreasing the pressure drop during withdrawal, thereby reducing not only the compressor investment cost but also the compressor operating cost. Table 7-2 illustrates this principle. In making this comparison it must be kept in mind that the additional well will probably not be as effective as the present wells because the new well may overlap some on the drainage pattern. Even if the additional well is only 50 percent effective and it decreases compressor fuel, the trade-off can be evaluated on an annual cost basis.

Table 7-2 is an example of a reservoir that has been proposed for development with four wells, and the characteristics of each well have been determined. The four identical wells give the reservoir a deliverability of 30,000 Mcf per day when the reservoir pressure is 1,150 psia and the flowing wellhead pressure is 285 psia. The flow rate of each well is 7,500 Mcf per day. It is desired to see what the economic effect of adding one more well will be. The total flow rate will be kept the same, and the pressure drop through the wells will be decreased. Since the originally proposed four wells are presumed adequate to drain the reservoir effectively, any additional well that is added will probably not be as effective as the original four. For this particular case, assume that the additional well is only half as effective as the original wells. This would have the effect of adding only half a well, although the cost would be the same as a full well.

With the additional well, the flow rate per effective well will be reduced from 7,500 Mcf per day to 6,667. With the reservoir pressure remaining

Table 7-2
Economic Trade-Off Between Wells and Withdrawal Compressors

	Base Case	One Extra Well
Number of Wells	4	5
Effective Number of Wells	4	4.5
Well C Factor	0.1	0.1
Well n Factor	0.8	0.8
Total Withdrawal Deliverability – Mcfd	30,000	30,000
Deliverability per Effective Well	7,500	6,667
Reservoir Pressure – psia	1,150	1,150
Wellhead Flowing Pressure – psia	285	501
Compressor Suct. Pressure – psia	260	476
Compressor Disch. Pressure – psia	800	800
Horsepower Required	2,115	909
Investment per Horsepower	2,000	2,000
Investment per Well	400,000	400,000
Investment Fixed Charges – Percent	25	25
Compressor O & M per Horsepower	220	220
Investment		
Wells	$1,600,000	$2,000,000
Extra Gathering System	0	$75,000
Compressors	$4,229,857	$1,817,554
Total	$5,829,857	$3,892,554
Cost of Service		
Wells	$400,000	$500,000
Extra Gathering System	0	$18,750
Compressors	$1,522,748	$654,320
Total	$1,922,748	$1,173,070

the same, this reduced flow rate will reduce the pressure drop in the well tubing and raise the flowing wellhead pressure from 285 psia to 501 psia. This in turn raises the suction pressure on the compressor from 260 psia to 476 psia. Keeping the same compressor discharge pressure, the compressor horsepower is reduced from 2,115 horsepower to 909 horsepower. The extra cost of the well is assumed to be $400,000 and the extra gathering system is $75,000. This is offset by the decreased cost of the compressors of $2,412,303. The annual cost of service is similarly reduced. Thus adding another well is economically sound, even though its effectiveness in draining the reservoir is not particularly good. It should be kept in mind that this example has dealt only with the withdrawal portion of the facility. A check would have to be made to make sure that the injection requirements are still met.

Working Gas Volume

It is important to be aware of the effect that each of the factors has on facility economics. One of the cases to be considered is the effect that the working gas volume has on the wellhead flowing pressure limitations. Assume that a reservoir has been selected that has a given volume, in this case 26.0 Bcf. The maximum reservoir pressure is 1,365 psia. Figure 7-1 shows a typical pressure decline curve for this type of reservoir. As the gas in the reservoir is produced, the reservoir pressure decreases. When the reservoir is full, the pressure is 1,365 psia. When all the gas has been removed from the reservoir, the pressure is 0.

As more of the total reservoir capacity is dedicated to working gas, the lower the reservoir pressure will be at the end of the withdrawal season. Figure 7-2 shows this effect. If the storage facility dedicates 6 of the 26 Bcf total capacity to working gas, the reservoir pressure when all the working gas has been withdrawn will be about 1,060 psia. If the working gas dedication is increased to 12 Bcf, the pressure at the end of the withdrawal season will be about 780 psia. Thus the amount of working gas determines the reservoir pressure at the end of the withdrawal season. As the reservoir pressure decreases, the flowing wellhead pressure must also decrease in order to maintain a given flow rate. The amount of working gas then determines the maximum flowing wellhead pressure that can be maintained at the end of the season in order to sustain a desired flow rate.

Figure 7-3 shows this effect graphically. The maximum flowing wellhead pressure at the end of the withdrawal season is plotted versus the amount of the reservoir that is dedicated to working gas. The parameters are rates of flow at the end of the season. The curve for 0 flow rate is simply the pressure decline curve for the reservoir. With a working gas dedication of 5 Bcf and a flow rate of 100 MMcf per day, the maximum flowing wellhead pressure is about 990 psia. Any pressure higher than this will not give the desired flow rate. If the working gas dedication is increased to 15 Bcf and the same 100 MMcf per day is desired, the flowing wellhead pressure must be about 205 psia or lower. The curve drops off

Reservoir Pressure (psia)

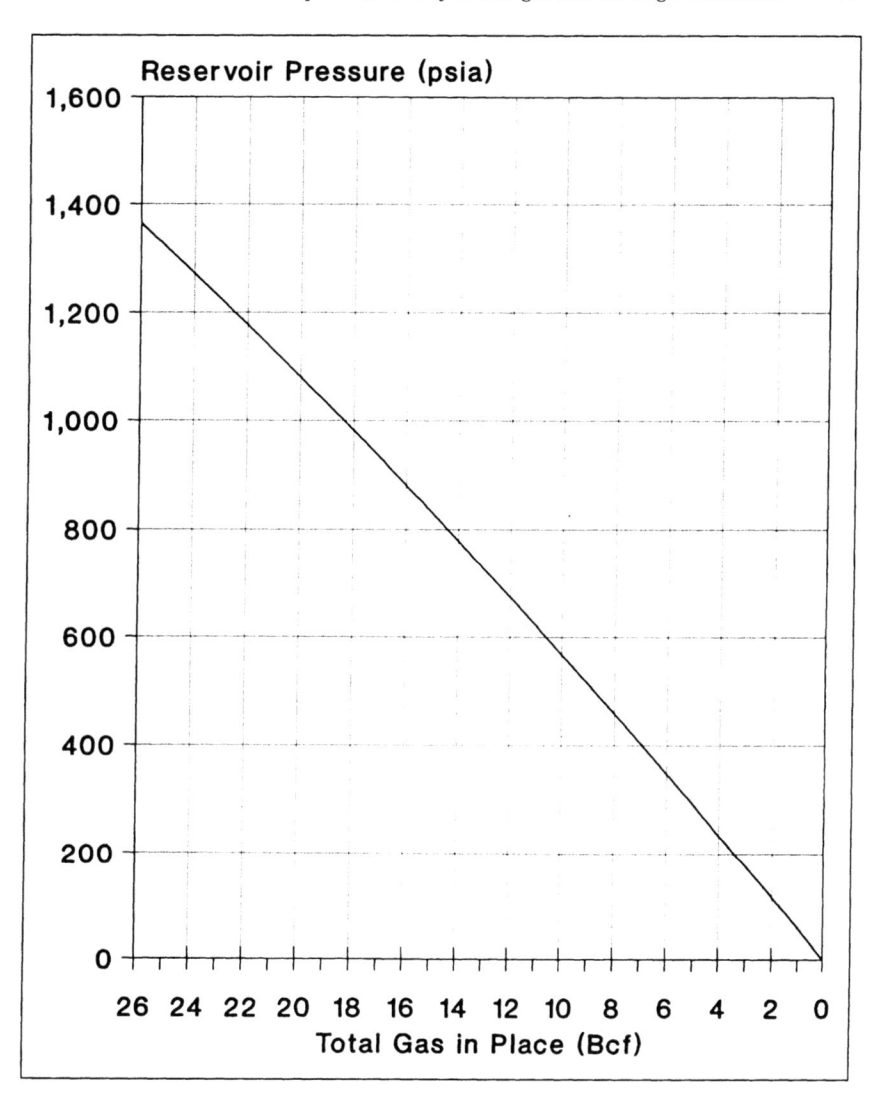

[Graph showing a linear decline from approximately 1,375 psia at 26 Bcf Total Gas in Place down to 0 psia at about 1 Bcf]

Total Gas in Place (Bcf)

Figure 7-1. Typical pressure decline curve for a volumetric reservoir.

sharply at the lower pressures because the flow rate is not linear with pressure drop. At low pressures the pressure drop must be higher than at high pressures to achieve the same flow rate.

The effect that the amount of working gas has on other parameters flows through to the withdrawal compressors. As the flowing wellhead pressure goes down, the pressure at the suction of the compressors goes down. The compressor suction pressure differs from the wellhead pressure by the amount of the pressure drop in the gathering system and the central point

74 Underground Gas Storage Facilities

Pressure at End of Withdrawal (psia)

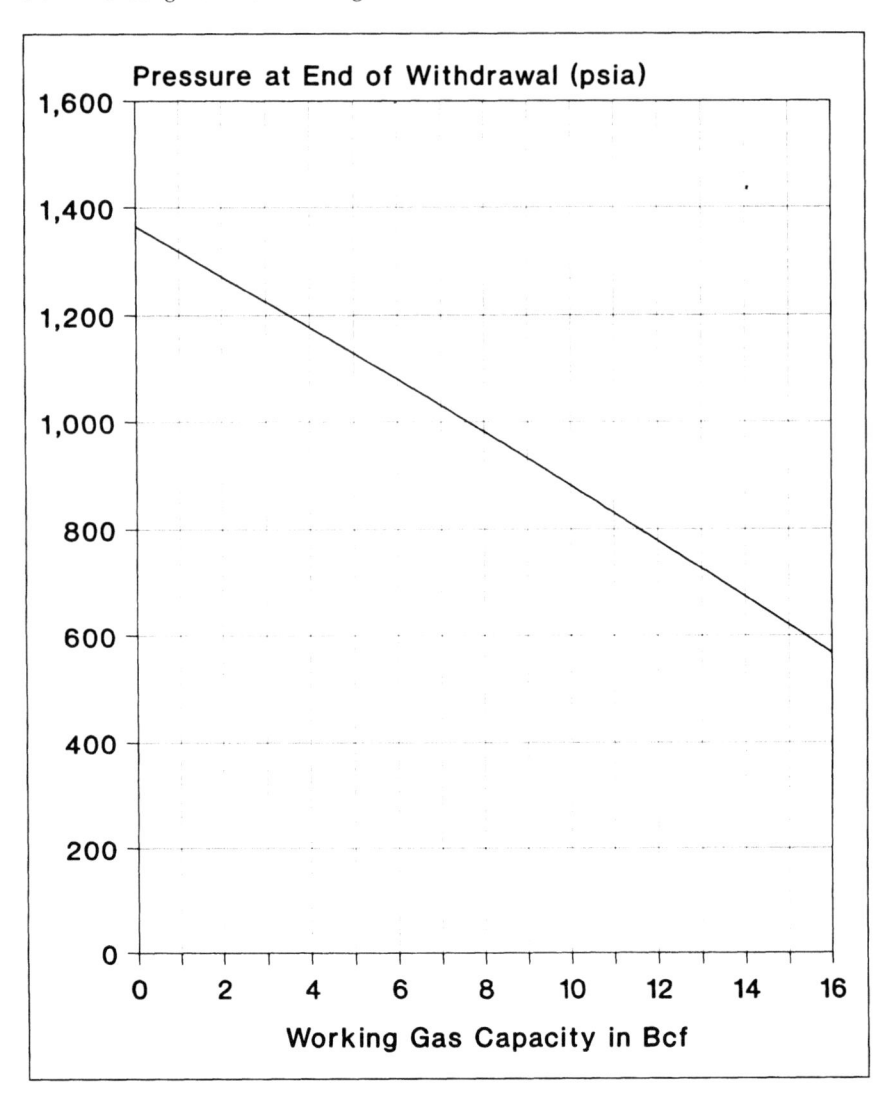

[Graph: x-axis "Working Gas Capacity in Bcf" from 0 to 16; y-axis from 0 to 1,600. A straight declining line from approximately (0, 1,360) to (16, 570).]

Figure 7-2. Effect of working gas capacity on reservoir pressure at end of withdrawal season.

facilities. Figure 7-4 takes the information from Figure 7-3 and applies it to the compressor installation. The x-axis is the amount of the reservoir dedicated to working gas. The right-hand y-axis is the compressor suction pressure at the end of the withdrawal season for a flow rate of 100 MMcf per day. The sharp drop-off in pressure that was noted in Figure 7-3 is not noted here, because the pressure was not plotted for the lower values where this occurred. The horsepower required for withdrawal is plotted on the

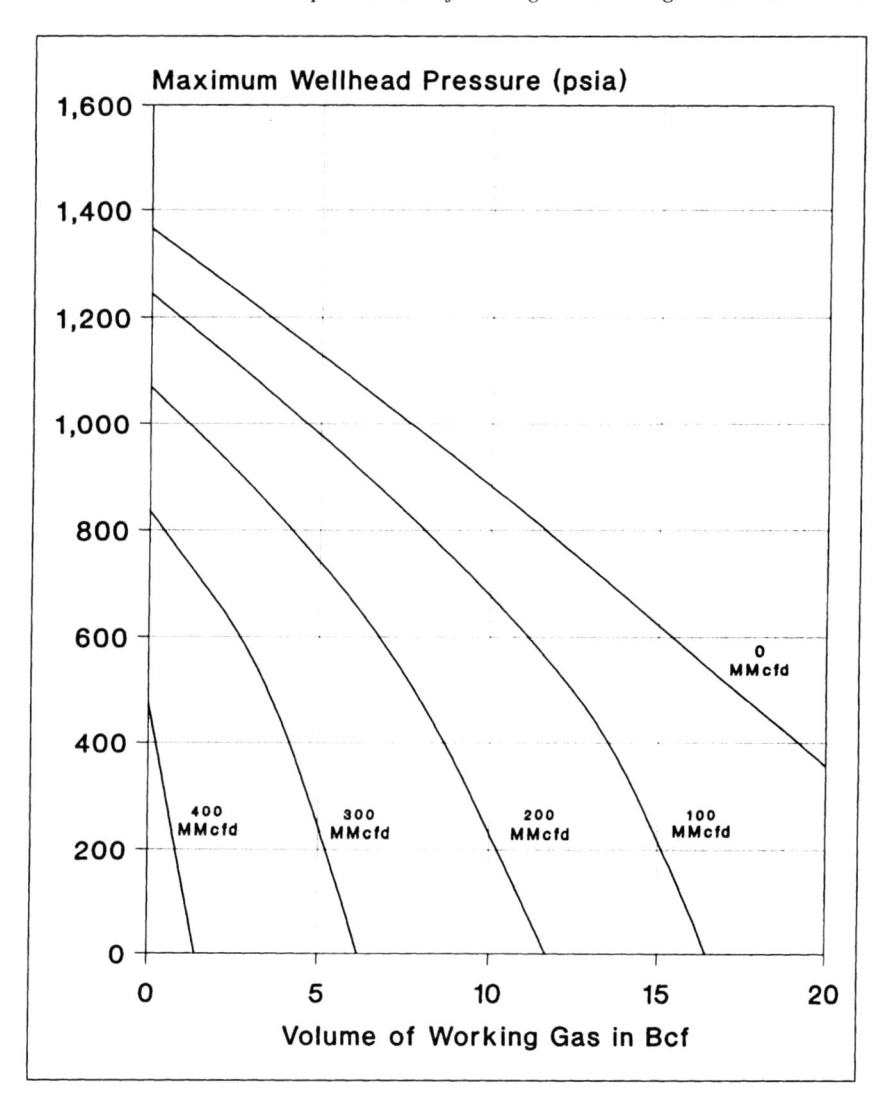

Figure 7-3. Effect of working gas volume on the maximum wellhead pressure.

left-hand y-axis. When the flow rate and the discharge pressure are held constant, the horsepower is a function of the suction pressure. As can be seen in Figure 7-4 this function is not linear. The horsepower increases faster than the suction pressure decreases. This is because the horsepower calculation contains the ratio of discharge pressure to suction pressure. This ratio is not linear with suction pressure. The result is that as the quantity of the reservoir that is dedicated to working gas increases, the horse-

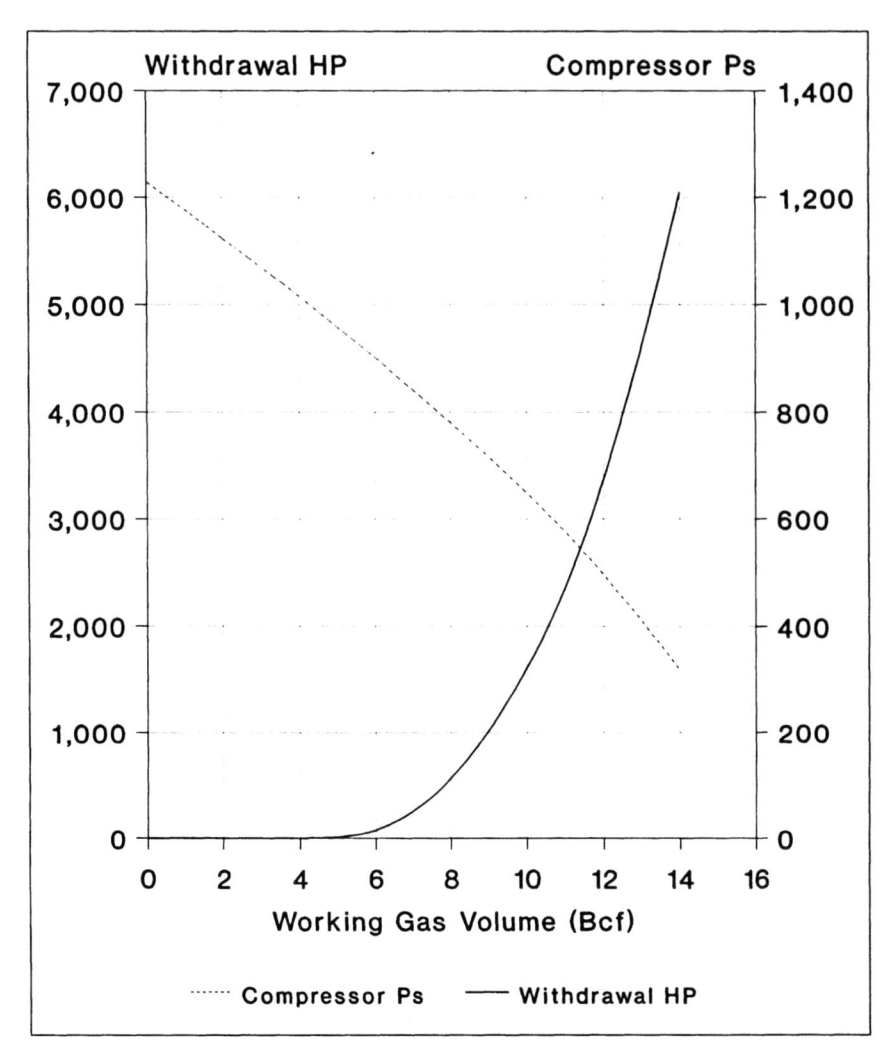

Figure 7-4. Effect of working gas volume on compressor functions for a compression rate of 100 MMcfd.

power required for withdrawal increases, and the increase is greater than the increase in working gas.

Injection Horsepower

There is no similar effect for the injection horsepower requirement. The injection horsepower is determined by the maximum reservoir pressure, or the reservoir pressure when it is full of gas. When the compressors are in injection service, they will be taking suction from the transmission system. This system pressure will be relatively stable, with somewhat random day-to-day variations. Thus the suction pressure on injection service will be relatively constant. When the working gas volume is divided by the number of injection days, the injection flow rate is determined. Therefore to this extent the volume of working gas affects the injection horsepower required. This effect by the injection flow rate is linear, however.

At the beginning of the injection season the pressure in the reservoir is relatively low, as all the working gas is gone. The injection compressors pump a constant volume against this low pressure, and a relatively small amount of horsepower is required. As the reservoir fills up with working gas, the reservoir pressure increases. This causes the compressor discharge pressure to increase, and this in turn causes the compressor horsepower required to increase. When all of the working gas has been injected, the reservoir is full and the reservoir pressure is at its maximum. The maximum injection horsepower is required at this point.

Types of Storage

One of the first decisions to be made in a storage development project is what type of storage is desired. An earlier chapter discussed the types of storage. These are base load, peak shaving, and mid-range types of storage. Each of these types can be further defined by the amount of time it takes to empty the working gas at the rated deliverability of the storage. The rated deliverability of the storage is the deliverability of the facility for the last Mcf of working gas. Table 7-3 shows the interaction of the various reservoir characteristics. The reservoir has a maximum capacity of 26.0 Bcf and a maximum pressure of 1365 psia. The first lines of Table 7-3 pertain to a wellhead flowing pressure of 800 psia. With this quantity fixed, the amount of working gas is varied. This affects the reservoir pressure at the end of the withdrawal season and thereby the deliverability at the end of the season. This also affects the time to empty the working gas at the rated deliverability. With 4 Bcf of working gas, this case would have a rating of 206 MMcfd, and the time to empty the working gas at this rating would be 19 days. By the definitions established in an earlier chapter, this would be classified as a peak shaving type of storage. With 10 Bcf of working gas and the same 800 psia flowing wellhead pressure, the rated deliverability would be 49 MMcfd, and the time to empty the working gas at this rating would be 203 days. This would definitely be a base load storage. It is questionable, however, whether this facility would be viable with

(text continued on page 79)

Table 7-3
Interaction of Reservoir Characteristics

Total GIP Bcf	Working Gas Removed Bcf	Reservoir Pressure psia	Wellhead Flowing Pressure psia	Deliverability Rating MMScfd	Time to Empty at Rating In Days
26	0	1,365	800	312.4	0
24	2	1,272	800	258.6	8
22	4	1,177	800	205.8	19
20	6	1,080	800	153.7	39
18	8	982	800	101.9	78
16	10	882	800	49.3	203
26	0	1,365	600	371.7	0
24	2	1,272	600	319.8	6
22	4	1,177	600	269.2	15
20	6	1,080	600	220.1	27
18	8	982	600	172.3	46
16	10	882	600	126.0	79
14	12	779	600	81.0	148
12	14	675	600	36.2	386
26	0	1,365	400	413.0	0
24	2	1,272	400	362.2	6
22	4	1,177	400	312.9	13
20	6	1,080	400	265.2	23
18	8	982	400	219.3	36
16	10	882	400	175.4	57
14	12	779	400	133.6	90
12	14	675	400	94.1	149
26	0	1,365	200	437.4	0
24	2	1,272	200	387.2	5
22	4	1,177	200	338.5	12
20	6	1,080	200	291.6	21
18	8	982	200	246.6	32
16	10	882	200	203.8	49
14	12	779	200	163.3	73
12	14	675	200	125.5	112
10	16	568	200	90.8	176
26	0	1,365	100	443.5	0
24	2	1,272	100	393.4	5
22	4	1,177	100	344.9	12
20	6	1,080	100	298.2	20
18	8	982	100	253.4	32
16	10	882	100	210.7	47
14	12	779	100	170.6	70
12	14	675	100	133.2	105
10	16	568	100	98.8	162

(text continued from page 77)

an empty time of 203 days. The remainder of Table 7-3 shows the effect of different flowing wellhead pressures on these same factors.

Table 7-4 illustrates this same interaction of the various features of a reservoir in a different format. This table shows the variation in deliverability at the end of the withdrawal season with working gas dedication for a given flowing wellhead pressure. For a working gas capacity of 6.0 Bcf and a flowing wellhead pressure of 600 psia, the rated deliverability at the end of the withdrawal season would be 220 MMcfd. If the wellhead flowing pressure were reduced to 400 psia for the same working gas, the deliverability rating would be increased to 265 MMcfd.

Table 7-4
Interaction of Reservoir Characteristics

Total GIP Bcf	Working Gas Removed Bcf	Reservoir Pressure psia	Deliverability Rating for the Following Wellhead Flowing Pressure (psia)					
			1000	900	800	700	600	500
26	0	1,365	232.8	275.3	312.4	344.4	371.7	394.5
24	2	1,272	175.6	220.1	258.6	291.7	319.8	343.3
22	4	1,177	118.0	165.4	205.8	240.2	269.2	293.4
20	6	1,080	58.4	110.7	153.7	189.8	220.1	245.1
18	8	982	–	54.5	101.9	140.5	172.3	198.4
16	10	882	–	–	49.3	92.0	126.0	153.6
14	12	779	–	–	–	43.1	81.0	110.5
12	14	675	–	–	–	–	36.2	69.3
10	16	568	–	–	–	–	–	28.9
8	18	459	–	–	–	–	–	–

Total GIP Bcf	Working Gas Removed Bcf	Reservoir Pressure psia	Deliverability Rating for the Following Wellhead Flowing Pressure (psia)					
			450	400	350	300	250	200
26	0	1,365	404.3	413.0	420.6	427.3	432.8	437.4
24	2	1,272	353.3	362.2	370.1	376.8	382.5	387.2
22	4	1,177	303.7	312.9	321.0	327.9	333.8	338.5
20	6	1,080	255.7	265.2	273.5	280.7	286.7	291.6
18	8	982	209.5	219.3	227.9	235.3	241.5	246.6
16	10	882	165.2	175.4	184.4	192.0	198.5	203.8
14	12	779	122.8	133.6	143.0	151.1	157.8	163.3
12	14	675	82.6	94.1	104.1	112.7	119.8	125.5
10	16	568	44.1	57.0	67.9	77.0	84.6	90.8
8	18	459	4.7	21.4	34.1	44.4	52.7	59.4

Table 7-5
Economic Design Factors

Reservoir Size – Bcf	26.0
Reservoir Maximum Pressure – psia	1,365
Compressor Cost per Horsepower	$2,000
Well Cost per Well	$450,000
Well C Factor	0.1041
Well n Factor	0.8435
Pipeline Cost per Inch–Mile	$20,000
Cushion Gas Cost per Mcf	$2.25
Compressor O & M Cost per Horsepower (Excl. Fuel)	$100
Compressor Fuel – Cubic Feet per Horsepower Hour	9
Cost of Compressor Fuel per Mcf	$2.25
Depreciation Rate – Percent	4.00
Return & Income Tax – Percent of Depreciated Investment	19.50
Taxes & Insurance – Percent of Depreciated Investment	1.25
General & Administrative – Percent of Depreciated Investment	0.50

Optimization

Possibly the best way to illustrate the economics of a storage facility is by an example. Table 7-5 lists the economic and physical design factors for a typical storage facility. The reservoir size and the well flow constants may vary considerably from case to case, but the reservoir properties shown here are reasonable. The economic factors will vary by company and with time. Therefore it is stressed that the value of each item is of less importance than the concept that it represents. The reservoir has a maximum capacity of 26.0 Bcf and a maximum pressure of 1365 psia. The compressor horsepower is priced at $2,000 per horsepower installed but exclusive of interest during construction and contingency. The wells and pipeline facilities are priced on a similar basis. The pipeline cost is stated on a per inch-mile basis. A nominal 12-inch pipeline that is 2 miles long contains 24 inch-miles and would be priced at $480,000.

Storage Emptying Time. The example will be for a base load storage facility, and the time to empty will be specified as 80 days. This specification is somewhat of a judgement call, as cases can be made for 70 days or 90 days. Table 7-6 shows the characteristics of the reservoir and wells for an empty time of 80 days. The values in this table were arrived at by

(text continued on page 82)

Table 7-6
Reservoir Choices for an Empty Time of 80 Days

Wellhead Pressure psia	Reservoir Pressure psia	Rated Flow Rate MMcfd	Total GIP Bcf	Working Gas Removed Bcf	Time to Empty in Days
1000	1,099	70.2	20.39	5.61	80.0
950	1,068	78.3	19.74	6.26	80.0
900	1,037	86.1	19.11	6.89	80.0
850	1,008	93.5	18.52	7.48	80.0
800	979	100.6	17.95	8.05	80.0
775	966	104.0	17.68	8.32	80.0
750	953	107.4	17.41	8.59	80.0
725	940	110.6	17.15	8.85	80.0
700	927	113.7	16.90	9.10	80.0
675	915	116.8	16.66	9.34	80.0
650	903	119.8	16.42	9.58	80.0
625	891	122.6	16.19	9.81	80.0
600	880	125.4	15.97	10.03	80.0
575	869	128.0	15.76	10.24	80.0
550	859	130.6	15.55	10.45	80.0
525	849	133.1	15.36	10.64	80.0
500	839	135.4	15.17	10.83	80.0
475	830	137.7	14.99	11.01	80.0
450	821	139.8	14.81	11.19	80.0
425	813	141.9	14.65	11.35	80.0
400	805	143.8	14.50	11.50	80.0
375	797	145.6	14.35	11.65	80.0
350	790	147.3	14.21	11.79	80.0
325	784	148.9	14.09	11.91	80.0
300	777	150.4	13.97	12.03	80.0
275	772	151.8	13.86	12.14	80.0
250	766	153.0	13.76	12.24	80.0
225	762	154.2	13.67	12.33	80.0
200	758	155.2	13.58	12.42	80.0
175	754	156.1	13.51	12.49	80.0
150	750	156.9	13.45	12.55	80.0
125	748	157.6	13.39	12.61	80.0
100	745	158.1	13.35	12.65	80.0

82 Underground Gas Storage Facilities

(text continued from page 80)

an iterative process. The wellhead pressures were specified. The reservoir pressures were estimated. It should be emphasized that the estimate of reservoir pressure does not need to be accurate, as the iterative process will refine them. The flow rate for the facility is calculated in the third column assuming 22 wells. The fourth column was arrived at by calculation. The estimated reservoir pressure was used to calculate the remaining total gas in place. This was done with the equation developed in Chapter 4:

$$GIP = a + \frac{b \times P}{c + d \times P} \quad (7\text{-}1)$$

P = reservoir pressure in psia
a = constant
b = constant
c = constant
d = constant

The constants used are those determined in Appendix A. Once the total gas in place is determined by this calculation, this gas in place is subtracted from the maximum reservoir capacity (26.0 Bcf) in order to arrive at the working gas. The volume of working gas is then divided by the rated flow rate in column three to determine the time to empty. The calculated time to empty will not be the 80 days desired, because the estimated reservoir pressure in column two is incorrect. The value in column two therefore needs to be corrected. Some algorithm needs to be fashioned to use the calculated days to empty to modify the assumed reservoir pressure. The equation used in this case is:

$$P_{n+1} = P_n + 1.5 \times [ET_{des} - ET_{cal}] \quad (7\text{-}2)$$

P_{n+1} = pressure trial for next iteration
P_n = pressure trial for last iteration
ET_{des} = desired empty time
ET_{cal} = empty time calculated by last iteration

This approach gives a convergence within a reasonable time.

Flowing Wellhead Pressure. Table 7-6 shows that a variety of flowing wellhead pressures will give an empty time of 80 days. This same information is also shown in Figure 7-5. The first approach should be to check the working gas volumes to see if there are a significant number of choices within the area of interest. Assuming that all of the values of working gas are viable, the choice then becomes an economic one.

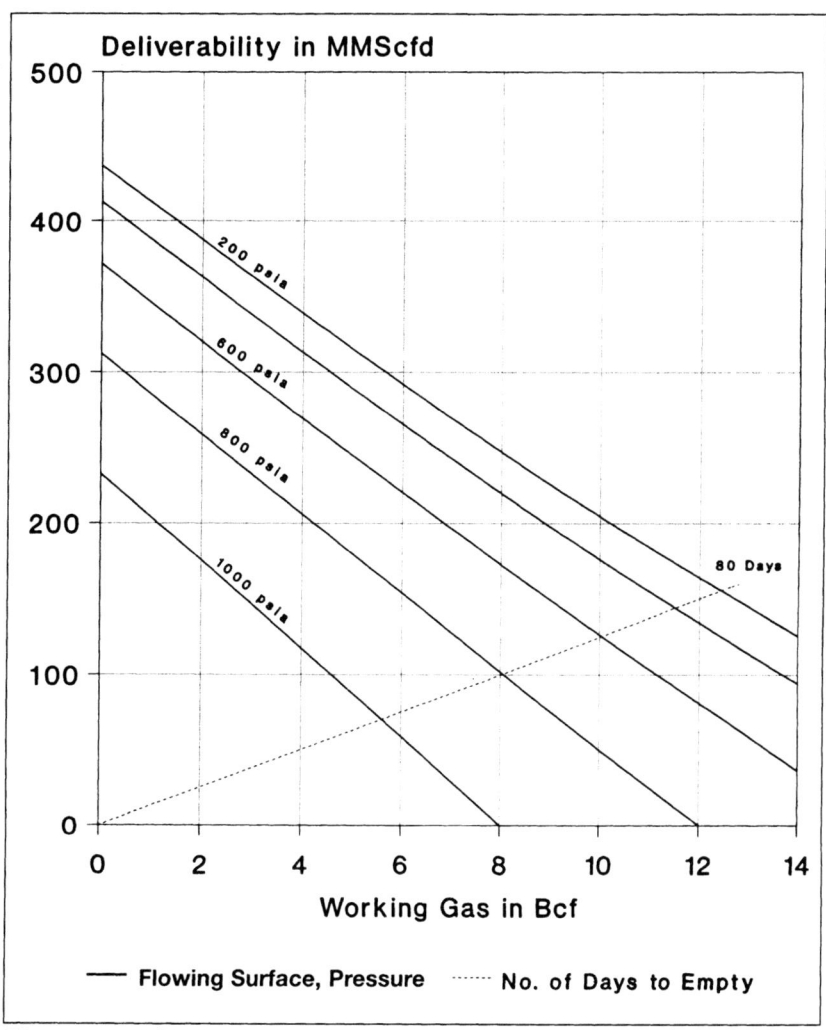

Figure 7-5. Characteristics of a storage with an 80-day withdrawal time.

Economics. Table 7-7 shows in tabular form how the economic choices are compared. The flowing wellhead pressure is selected as the primary variable, and the other factors react to this. Due to space limitations, only 5 columns of pressure choices are shown. These are 200, 400, 600, 800, and 1,000 and cover the range of interest, if in somewhat large steps. A large number of rows are used to show the approach that was taken in the economic evaluation. The deliverability rating and the working gas

Table 7-7
Economics of Storage Facility Development Basis: 80 Days to Empty Storage at Rated Deliverability

	Wellhead Flowing Pressure – psia				
	200	400	600	800	1000
Deliverability Rating – MMcfd	155.2	143.8	125.4	100.6	70.2
Working Gas – Bcf	12.4	11.5	10.0	8.1	5.6
Cushion Gas – Bcf	13.6	14.5	16.0	17.9	20.4
Number of Wells	22	22	22	22	22
Injection Rate – MMcfd	69.0	63.9	55.7	44.7	31.2
Injection Suction Pressure – psig	530	524	519	513	510
Injection Discharge Pressure – psig	1,489	1,468	1,447	1,426	1,415
Withdrawal Suction Pressure – psig	170	371	573	776	980
Withdrawal Discharge Pressure – psig	780	765	857	815	840
Injection Horsepower Required	4,425	4,084	3,547	2,835	1,974
Withdrawal Horsepower Required	15,563	6,222	2,904	272	0
Horsepower Installed	20,751	8,295	4,729	3,781	2,631
Pipeline Size – Inches	16	16	12	12	10
Pipeline Length – Miles	12	12	12	12	12
Investment in Thousands of Dollars					
Compressors – Operating	31,126	12,443	7,093	5,671	3,947
Compressors – Spare	10,375	4,148	2,364	1,890	1,316
Wells	9,900	9,900	9,900	9,900	9,900
Gathering System	1,380	1,380	1,200	1,200	1,200
Transmission Line	3,840	3,840	2,880	2,880	2,400
Meters & Dehydrators	900	825	750	675	600
Interest, Contingency, etc.	5,752	3,254	2,419	2,222	1,936
Cushion Gas	30,562	32,617	35,933	40,387	45,867
Total	93,837	68,407	62,540	64,824	67,166
Cost of Service in Thousands of Dollars					
Return & Income Tax	18,051	13,200	12,091	12,545	13,014
Compressor O & M	2,075	830	473	378	263
Compressor Fuel	1,656	662	377	302	210
Other O & M	250	250	250	250	250
Depreciation	2,531	1,432	1,064	978	852
Taxes & Insurance	1,157	846	775	804	834
General & Administrative	463	338	310	322	334
Total	26,184	17,558	15,341	15,579	15,757
Working gas – MMcf	12,417	11,503	10,030	8,050	5,615
Cost of Service per Mcf	2.11	1.53	1.53	1.94	2.81

volume were taken from Table 7-6. The cushion gas capacity is obtained by subtracting the working gas volume from the reservoir capacity. The number of wells is set at 22. An injection period of 180 days is assumed, and the injection rate is obtained by dividing the working gas volume by this value.

The gathering system and the transmission line should be designed for each of these cases. The only variation from case to case should be in the size of the transmission line and in the size of some of the trunk lines in the gathering system. This variation in size will be due to the variation in withdrawal rates. The entire facility should be simulated by a computer model. This computer model will then calculate the pressures at the various points in the system and also calculate the required horsepower. This simulation should be made for both the injection cycle and the withdrawal cycle. The cycle that requires the most horsepower will then govern. It should be noted that there is a shift in the governing cycle as the pressure changes. The computer model (or a manual calculation) will determine the minimum required horsepower. Since horsepower comes in size blocks, it is not always possible to purchase the minimum horsepower required. The available sizes must be tailored to the needs. This usually results in an increase above the minimum required. In addition, some spare horsepower is required. The amount of this spare horsepower will depend on the operating philosophy of the individual company. Quite often a single spare compressor is purchased. If the sizing is such that three units are required, four are purchased. In Table 7-7 the installed horsepower includes 33 percent spare capacity.

The investment costs are calculated using the economic factors in Table 7-5. For the cases of flowing wellhead pressures of 200 and 400 psia the withdrawal compressor horsepowers govern. For the cases of 600, 800, and 1,000 psia the injection compressor horsepowers govern. The investment in compressors is broken down into the operating units and the spare unit. No attempt was made to block out the compressor horsepower into commercially available units. The wells, gathering system, and the transmission also were priced from the design using the economic factors in Table 7-5. The meters and dehydrators were priced as installed units. In order to allow for interest during construction and contingencies, a factor of 10 percent of these costs was added. The cushion gas was priced at $2.25 per Mcf. No interest during construction or contingency was added to the cushion gas. It can be seen that in most of the cases the cushion gas is the largest single portion of the investment.

The annual cost of service was calculated using the investment and the factors in Table 7-5. The allowance for return and income tax was applied to the total depreciated investment. The cost of service will vary from year to year because of this. The numbers shown are for the first year. These figures are based on an average depreciation of one half year. The compressor operating and maintenance costs are applied to all the compressors. The compressor fuel costs are applied to only the operating compressors and assume that they operate an average of 60 percent during the year. Operating and maintenance costs for the pipelines, dehydrators, and other equipment is set at a flat figure of $250,000 per year.

86 Underground Gas Storage Facilities

Optimization Procedure. The total annual cost of service is divided by the total working gas to determine the cost of service per Mcf of working gas. It can be seen that an optimum point lies between a flowing wellhead pressure of 400 psia and 600 psia. It can also be seen that the cost variation per Mcf of working gas is quite large.

Table 7-8 shows the same type of information but in a different format. Much less detail is shown as to the procedure that was used in calculating the economics, but a much more detailed examination of the pressure ranges is shown. The same procedure was used as was displayed in Table 7-7, and the end result for the same wellhead pressures is the same. Table 7-7 showed that the optimum in terms of cost of service per Mcf of working gas occurred between flowing wellhead pressures of 400 psia and 600 psia. With Table 7-8 this range can be examined in more detail. The optimum occurs between 475 and 500 psia, probably about 480 psia. Inter-

Table 7-8
Economics of Storage Facility Development Basis: 80 Days to Empty Storage at Rated Deliverability

Well-head Press. psia	Rated Flow Rate MMcfd	Installed Horse-power	Base Project Cost $1,000	Cost of Horse-power $1,000	Cost of Cushion Gas $1,000	Total Project Cost $1,000	Annual Cost of Service $1,000	Working Gas Volume Bcf	Cost of Service per Mcf
1000	70.2	2,631	15,510	5,789	45,867	67,166	15,757	5.61	2.81
950	78.3	2,937	15,510	6,462	44,405	66,377	15,669	6.26	2.50
900	86.1	3,231	15,510	7,108	43,004	65,622	15,584	6.89	2.26
850	93.5	3,512	15,510	7,727	41,664	64,901	15,504	7.48	2.07
800	100.6	3,781	16,121	8,317	40,387	64,824	15,579	8.05	1.94
775	104.0	3,911	16,121	8,604	39,772	64,496	15,543	8.32	1.87
750	107.4	4,038	16,121	8,883	39,173	64,177	15,508	8.59	1.81
725	110.6	4,162	16,121	9,156	38,591	63,867	15,474	8.85	1.75
700	113.7	4,282	16,121	9,420	38,025	63,566	15,441	9.10	1.70
675	116.8	4,399	16,121	9,678	37,476	63,275	15,409	9.34	1.65
650	119.8	4,513	16,121	9,928	36,945	62,993	15,379	9.58	1.61
625	122.6	4,623	16,121	10,170	36,430	62,720	15,349	9.81	1.56
600	125.4	4,729	16,203	10,404	35,933	62,540	15,341	10.03	1.53
575	128.0	4,832	16,401	10,630	35,454	62,485	15,363	10.24	1.50
550	130.6	5,145	16,401	11,318	34,993	62,712	15,492	10.45	1.48
525	133.1	5,855	16,401	12,880	34,550	63,831	15,914	10.64	1.50
500	135.4	5,118	17,457	11,259	34,125	62,841	15,551	10.83	1.44
475	137.7	5,696	17,457	12,532	33,719	63,708	15,884	11.01	1.44
450	139.8	6,491	17,457	14,280	33,333	65,069	16,379	11.19	1.46
425	141.9	7,354	17,457	16,180	32,965	66,602	16,928	11.35	1.49
400	143.8	8,295	17,540	18,250	32,617	68,407	17,558	11.50	1.53
375	145.6	9,324	17,540	20,514	32,289	70,342	18,235	11.65	1.57
350	147.3	10,454	17,540	22,998	31,981	72,519	18,989	11.79	1.61
325	148.9	11,700	17,540	25,741	31,693	74,973	19,833	11.91	1.66
300	150.4	13,085	17,540	28,787	31,425	77,751	20,781	12.03	1.73
275	151.8	14,636	17,540	32,198	31,178	80,916	21,855	12.14	1.80
250	153.0	16,391	17,540	36,059	30,952	84,551	23,081	12.24	1.89
225	154.2	18,403	17,540	40,487	30,747	88,773	24,498	12.33	1.99
200	155.2	20,751	17,622	45,652	30,562	93,837	26,184	12.42	2.11
175	156.1	23,553	17,622	51,816	30,400	99,837	28,183	12.49	2.26
150	156.9	27,002	17,622	59,404	30,258	107,285	30,657	12.55	2.44
125	157.6	31,445	17,622	69,179	30,138	116,940	33,857	12.61	2.69
100	158.1	37,587	17,622	82,692	30,040	130,355	38,296	12.65	3.03

estingly enough, the minimum investment occurs at a slightly different point, around 575 psia. At a flowing wellhead pressure of 475 psia there would be a working gas volume of 11.0 Bcf and a deliverability rating at the end of the season of 138 MMcfd. The total investment for the facility would be 64 million dollars, with over half of this cost being cushion gas. The annual cost of service for the storage would be $16 million per year, or $1.44 per Mcf of working gas.

Figure 7-6 shows graphically the same result of varying the wellhead flowing pressure. The optimum point appears to be around 480 to 490 psia. There is a "wiggle" in the curve around 525 to 550 psia. This is caused by a step change in the investment cost for the gathering system and the transmission line which resulted from changes in pipe sizes. The curve for Figure 7-6 is relatively flat, so that the wellhead flowing pressure can be varied significantly without greatly affecting the cost of service per Mcf. This is not always the case. In some instances the optimum point is quite acute and the cost of service per Mcf rises sharply on each side of the optimum. This usually occurs when the optimum wellhead flowing pressure is lower, perhaps 200 to 250 psia, and the increase in compressor horsepower becomes a large cost factor.

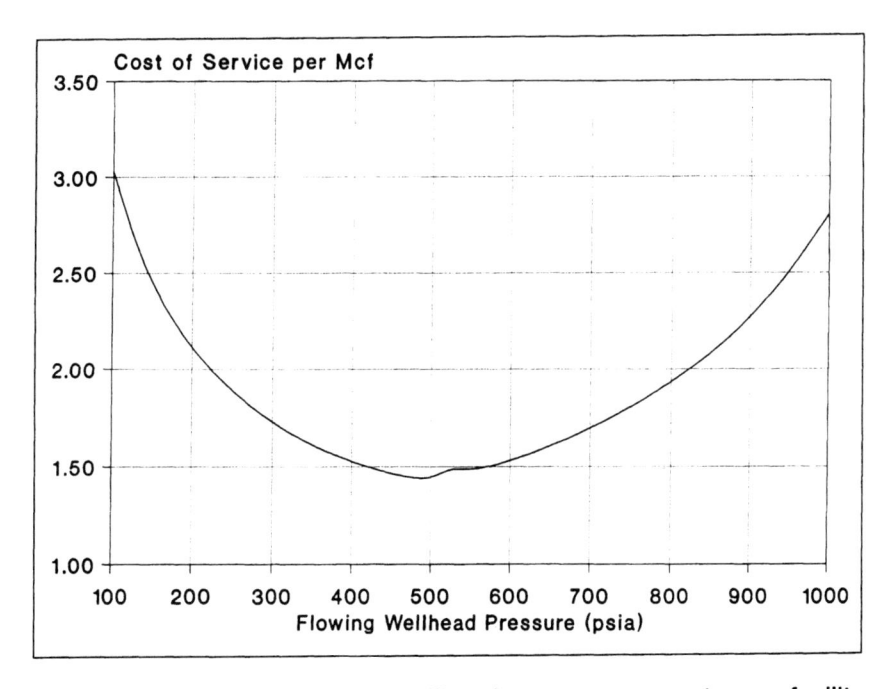

Figure 7-6. Effect of flowing wellhead pressure on a storage facility cost of service.

Table 7-9
The Economic Effect of Adding an Extra Well on the Cost of Withdrawal Compression

	Base Case	Effectiveness of Extra Well		
		0.50	0.75	1.00
Actual No. of Wells	22.00	23.00	23.00	23.00
Effective No. of Wells	22.00	22.50	22.75	23.00
SISP	830	830	830	830
FSP	475	487.6	493.6	499.4
C Factor	0.1041	0.1041	0.1041	0.1041
n Factor	0.8435	0.8435	0.8435	0.8435
Mcfd per Well	6,258	6,119	6,051	5,986
Total Mcfd	137,670	137,670	137,670	137,670
Comp. Suction Pressure	447.0	459.6	465.6	471.4
Comp. Discharge Pressure	759.8	759.8	759.8	759.8
Horsepower Required	4,272	4,034	3,926	3,822
Horsepower Installed	5,696	5,379	5,235	5,096
Investment in $1,000				
Well Cost	$9,900	$10,350	$10,350	$10,350
Extra Gathering System	0	75	75	75
Compressor Cost	11,393	10,758	10,470	10,192
Interest & Contingency	2,129	2,118	2,089	2,062
Total	$21,293	$21,183	$20,895	$20,617
Cost of Service in $1,000				
Wells	$2,945	$3,079	$3,079	$3,079
Extra Gathering System	0	22	22	22
Compressors	4,414	4,168	4,056	3,948
Interest & Contingency	633	630	622	613
Total	$7,992	$7,899	$7,779	$7,663
Annual Savings in $1,000	–	$93	$213	$329

Effect of Additional Wells. Table 7-8 represents the costs based on the number of wells recommended by the petroleum engineer. It would now be appropriate to see what the effect of adding one well would be. It should be remembered that the approach taken here is that the extra well is for the purpose of decreasing pressure drop, not increasing flow rate. Of course it will do either, but the economic effect being looked for here is in the pressure drop reduction. Table 7-9 shows this case. The base case is for the 22 wells and the flowing wellhead pressure of 475 psia. This was the optimum case found in Table 7-8. One additional well is added at a cost of $450,000. The first column showing the extra well assumes that the effectiveness of the extra well is only 50 percent. This means that the effect of the well is the same as adding half a well. With this assumption, the total flow rate and the reservoir pressure are held constant. The total flow rate is then divided by 22.5 in order to obtain the flow rate per well. This rate per well is now 6,119 Mcfd as opposed to 6,258 Mcfd for the 22 wells. This reduced flow rate reduces the pressure drop in the wells and raises the flowing wellhead pressure from 475 psia to 488 psia. This in turn raises the compressor suction pressure from 447 psia to 460 psia. This causes a reduction in required horsepower from 4,272 to 4,034.

The investment area in Table 7-9 shows the cost of the extra well and the additional gathering system to connect it. The horsepower investment cost is reduced, including the cost of spare horsepower. Interest and contingency is added on. There is an investment advantage of about $110,000 to adding the extra well. The annual cost of service savings is about $93,000. It appears that in this particular case the addition of one or more wells is worth investigating. The two right-hand columns in Table 7-9 show the effect of adding the extra well if the effectiveness of the added well is 75 percent and 100 percent, respectively. These values are not completely unreasonable. With the reduced flow rate the drainage effect of the original 22 wells would be decreased, so that it is conceivable that a relatively high percentage effectiveness could be achieved with the additional well or wells provided that they are designed into the well spacing pattern early in the project.

Summary

This chapter has brought out several points that should be summarized:

- The independent investigation of the use of compressors during the withdrawal cycle can result in savings in investment cost and in annual cost of service.

- The flowing wellhead pressure very definitely affects the investment and cost of service.

- There are optimum design points for investment cost and annual cost of service that are significantly lower than other design points. The optimum point for investment cost and the optimum point for cost of service are not necessarily the same.

- In some cases, it is cost-effective to add additional wells in order to reduce the compressor horsepower and the total project cost.

CHAPTER 8

Monitoring and Control of Inventory

Produced gas has a value. When this produced gas is put back into the ground in a storage reservoir, there is always concern about possible gas loss. Cushion gas and working gas together are the largest single cost item in a storage facility. The combination of the two will cost tens of millions and sometimes hundreds of millions of dollars. With this large exposure to risk it is natural to want to find some method of verifying the amount of gas in storage.

Gas is metered when it is put into storage, and it is metered when it is removed from storage. The difference between these two values plus the amount of native gas that was already in the reservoir make up the amount of gas that should be in storage. There are three factors that may cause this calculated value of inventory to differ from the actual amount of gas in storage:

- There was an error in the computation of the original amount of native gas remaining in the reservoir when injection was started. Although this is normally not a problem, it should be kept in mind when a discrepancy is detected.
- There is an error in the metering of the gas, either in to or out of the storage unit, or both. (Because metering is such an important topic, it is the subject of a separate chapter, Chapter 10.)
- Gas has been lost from the storage due to leakage.

Leakage is one of the major concerns of the petroleum engineer when a reservoir is examined as a potential storage unit. One of the major comforts in using a former producing reservoir as a storage facility is the fact that it has held gas before for long periods of time and therefore should be secure for use as a storage unit. While there is logic in this, the comfort that it engenders can sometimes be misplaced. It is true that the reservoir held gas securely for long periods of time and was quite secure under the conditions that existed at that time. When a gas field begins to be produced, there are changes that take place. One of the most common is that there is encroachment of water to fill up the space that was once occupied by gas. In some cases the invading water can completely fill the reservoir except for the small gas cap that is left at the end of production.

When this occurs, the engineer is faced with a reservoir that is essentially full of water. When the injection of gas begins in this situation, strange things can occur. The injected gas does not normally push down the water vertically in a stable, even manner, maintaining a horizontal continuous interface. There is a strong tendency for the gas to override the water in a horizontal direction. When the formation that originally held the gas is in the shape of a dome, the gas will move along the top of the dome. It will move along the dome roof in a horizontal downward direction until it reaches points where it is below the water level in the center of the dome. This type of behavior can cause two problems. First, as the gas stretches out in "fingers" through the water, the water can and does close in behind the gas. This establishes separate, independent pockets of gas that are cut off from the main body. Although these gas pockets may remain in the reservoir and are technically in inventory, they are essentially lost in that they cannot be produced without the production of large amounts of water.

The second problem is that the gas fingers may travel along the dome roof down so far that they reach a low hydrographic seal point for that portion of the reservoir. This type of seal is sometimes called a saddle because of its shape. When gas passes this seal, it is lost from the reservoir. There have been cases where large amounts of gas have been lost in this manner. For these reasons, the initial injection of gas into the reservoir should be done with a great deal of study. Sometimes an observation well is drilled near a saddle to permit the observation of the water in the structure. When the water fills the entire storage sand, it indicates that there is no gas at that point. When a gas-water interface is detected, it indicates that the gas has moved toward the observation well to a degree that corrective measures may need to be taken. A prospective reservoir that has filled with water presents a very severe challenge for the development into a storage facility. A much better choice would be a volumetric reservoir or a reservoir that has a less active water drive.

Types of Leakage

The following is a list of the ways that gas may be lost by leakage:

- Leaks around old well casings to other formations.
- Leaks through old well casings to other formations.
- Leaks through the cap rock.
- Leaks through a low permeability connection to a companion reservoir that is not part of the storage facility.
- Leaks through rock faults that may have moved as the pressure in the reservoir was lowered.
- Leaks due to gas traveling past a saddle seal.
- Leaks in surface equipment and pipelines.

The leakage of gas through the cap rock is not at all common. A more likely candidate is the existing wells in a storage field. Most all storage fields have the original producing wells still in place, and the storage facility may be using them for either withdrawal wells or for observation wells. If these wells are old, they may have been drilled at a time when the completion technology and reliability were not as good as they are today. In addition, if the wells are old, the cement and metal may have deteriorated. This type of situation can lead to the gas having a path from the storage reservoir up or down hole to another producing or nonproducing structure.

Pressure-Volume History

The leakage of gas from storage can often be detected through an examination of the pressure-volume history of the reservoir. In order to accomplish this, it is first necessary to understand what a normal history cycle looks like. The injection and withdrawal of gas from storage causes pressure changes in the reservoir. When injection and withdrawal cycles are identical from year to year and there is no leakage, the pressure-volume history should be identical from year to year. The injection and withdrawal cycles are never identical for two years in a row, but there should be points for comparison somewhere on the cycle.

Figure 7-1 in Chapter 7 is a pressure decline curve for a producing reservoir. The points along this curve assume complete pressure equalization throughout the reservoir for each pressure point. In order to use this type of information for storage service, it is convenient to plot it in a slightly different format. If the *x*-axis were changed from total gas produced to total gas remaining in place, the result would be a pressure-volume curve for the reservoir. This is shown in Figure 8-1. The dashed line in this figure is the pressure decline curve that has been reformatted. It is now a plot of reservoir pressure versus total gas in place.

If this reservoir were to be used for storage service, the working gas portion of the reservoir could be represented by the solid line in Figure 8-1. There would be about 15 Bcf of cushion gas and 11 Bcf of working gas.

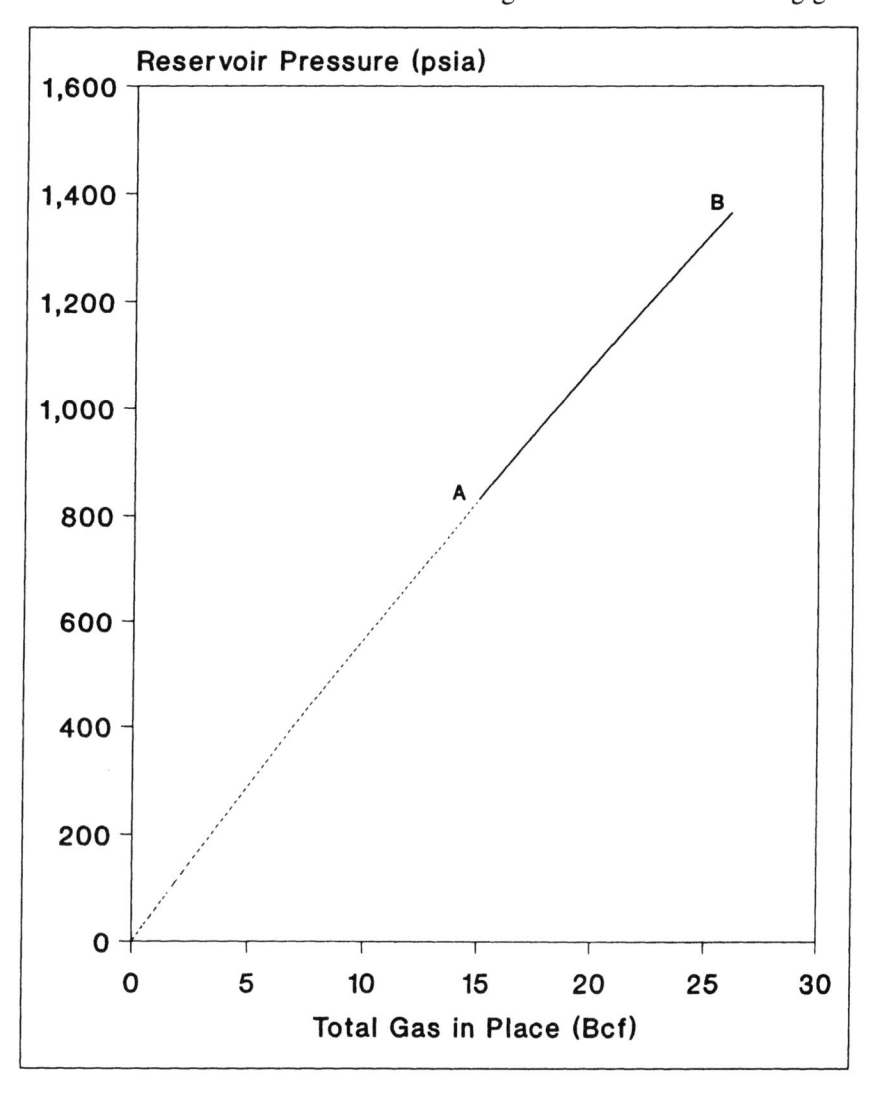

Figure 8-1. Idealized injection-withdrawal cycle for a volumetric reservoir.

The injection cycle would be from A to B and would occur over the summer months. During this time the reservoir would be filled to capacity. The withdrawal cycle would be from B to A. This would be a highly idealized type of operation. This type of operation could only occur with an extremely high permeability. A salt dome type of storage would fit this description. There would be no time delay or transients in the pressure profile of the reservoir. The pressures throughout the reservoir would be equal at all times. The storage operation would always be along the dotted line in Figure 8-1.

Volumetric Reservoir. A more realistic type of storage operation is shown in Figure 8-2. This figure is for a volumetric reservoir that has permeability values that are normally encountered in storage fields. The dashed line still represents the pressure decline curve for the reservoir. The injection period would be from A to B. At point B the reservoir is full and the pressure is significantly above the pressure decline curve. This is because the pressure is being measured at one or more of the injection-withdrawal wells. The pressure has not equalized throughout the reservoir, and the pressure at the well is higher than the rest of the reservoir. At the end of the injection cycle the storage field is usually shut in for a time period. This time period varies among various companies, but a typical time period is 15 to 30 days. One of the purposes of this shut-in period is to allow the pressures to equalize so that a check can be made on the gas inventory in storage.

This shut-in period is represented by the vertical line B to C in Figure 8-2. It can be seen that there is a significant drop in the pressure during this shut-in period. At the end of the period, the pressure is shown at point C. Point C is still above the pressure decline curve, indicating that the pressure is not completely equalized throughout the reservoir.

The withdrawal period is shown by the line C to D on the figure. As the gas is withdrawn, the pressure at the well drops below the pressure decline curve. This continues until point D. Point D shows the pressure in the well at the end of withdrawal. This pressure is considerably below the pressure decline curve, indicating that the pressure is far from being equalized in the reservoir. The usual custom is to have another shut-in period at the end of the withdrawal season. This period is represented by line D to A on the figure. During this shut-in period, the pressure rises from point D to point A. Although this is a significant pressure rise indicating that some pressure equalization has occurred, the pressure at point A is still below the pressure decline curve. This indicates that the reservoir pressure is still not completely equalized.

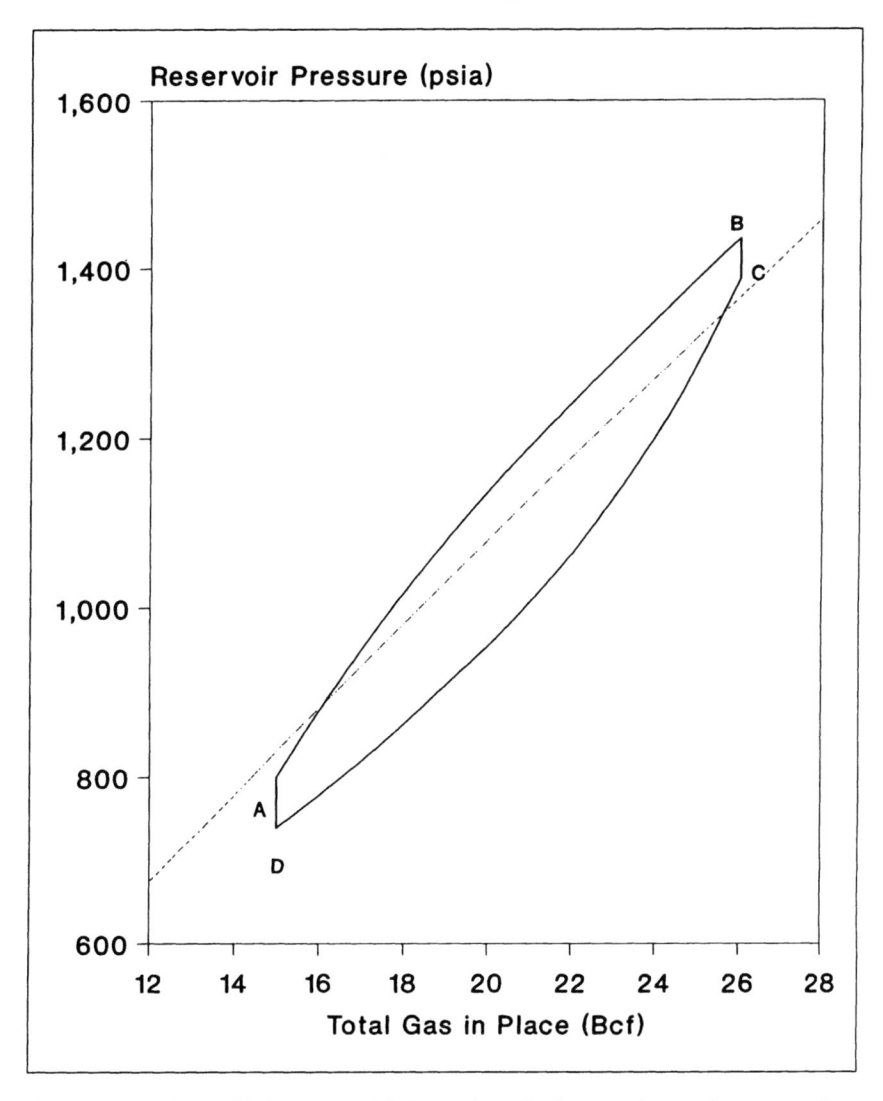

Figure 8-2. Typical injection-withdrawal cycle for a volumetric reservoir.

Water Drive Reservoir. Figure 8-2 illustrates that the behavior of a volumetric storage reservoir during the injection-withdrawal cycle is complex due to the transient nature of conditions in the reservoir. The behavior of a water drive reservoir in storage service is even more complex. Figure 8-1 showed the pressure decline curve for a volumetric reservoir. An aquifer or a reservoir with a strong water drive has a different type of pressure decline curve. This is illustrated by Figure 8-3.

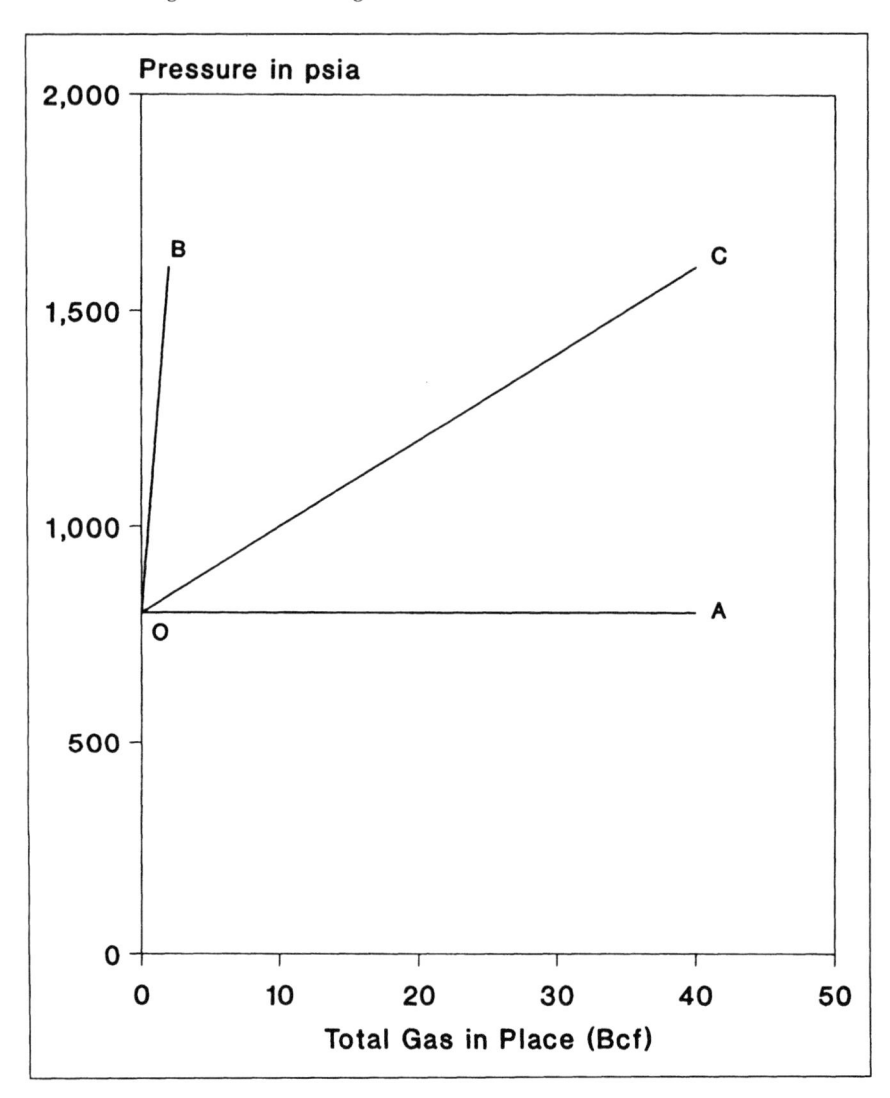

Figure 8-3. Pressure decline curves for some water drive reservoirs.

In a volumetric reservoir, when all the gas has been removed the pressure falls to 0. This is not necessarily true for a water drive reservoir. There is a pressure that is inherent with the water drive reservoir. It is dependent on the depth of the reservoir, and can be calculated if the depth is converted to a column of salt water. Point O in Figure 8-3 shows this point. If the reservoir were located at a deeper sub-surface elevation, point O would be at a higher pressure on the graph.

When a water drive reservoir is produced to depletion, the water encroaches into the reservoir and essentially fills it. When the field is converted to a storage facility and gas is injected, the injected gas pushes back the water and exposes more volume in the permeable sand to hold gas. This pushing back of the water is accomplished by two mechanisms. The first mechanism can be illustrated by turning a bucket upside down in a tank of water and introducing air into the bucket. This introduction of air pushes the water out and the air pressure in the bucket is balanced by the hydrostatic pressure of a column of water. The second mechanism is that the pressure of the gas actually compresses the water to a smaller volume. This can be thought of as a large volumetric reservoir containing a very large amount of water and a small amount of gas. Although the compressibility of the water is small, when the volume is large this compressibility results in significant volume change.

In Figure 8-3, the line OA represents a water drive reservoir that has an infinite volume of water. As gas is injected into the reservoir, the pressure change is not discernible. This case cannot be achieved in practice, and it may be thought of as a limiting case. The line OB represents another limiting case. Here the conditions are such that the injection of a small amount of gas causes an infinite rise in pressure. This is also not a feasible case. The pressure decline curves for water drive reservoirs lie between these two extremes. The line OC in Figure 8-3 illustrates a typical pressure decline curve for a reservoir with a strong water drive. It should be emphasized that all the lines in Figure 8-3 represent equalized pressure conditions in the reservoir.

Figure 8-4 shows the operating cycle for a typical water drive type of reservoir. This diagram looks quite similar to the operating cycle for the volumetric reservoir shown in Figure 8-2. There are some significant differences, however. In both cases the dashed line represents the pressure decline curve for the reservoir. In the volumetric case, however, this line passes through the origin of the graph. In the water drive case, it does not.

In Figure 8-4 the injection period is represented by line AB. The first portion of this line may be somewhat steeper than the same line in Figure 8-2. This is because the water has encroached during the low-pressure period and the gas is being injected into a smaller volume. This effect may not be discernable unless the reservoir has a high permeability. The high friction to gas flow around the well bore may prevent the gas from "seeing" that the volume is smaller. The pressure at point B is higher because the pressure has not entirely pushed the water back. Line BC shows the shut-in period after injection. During this period, the pressure drops con-

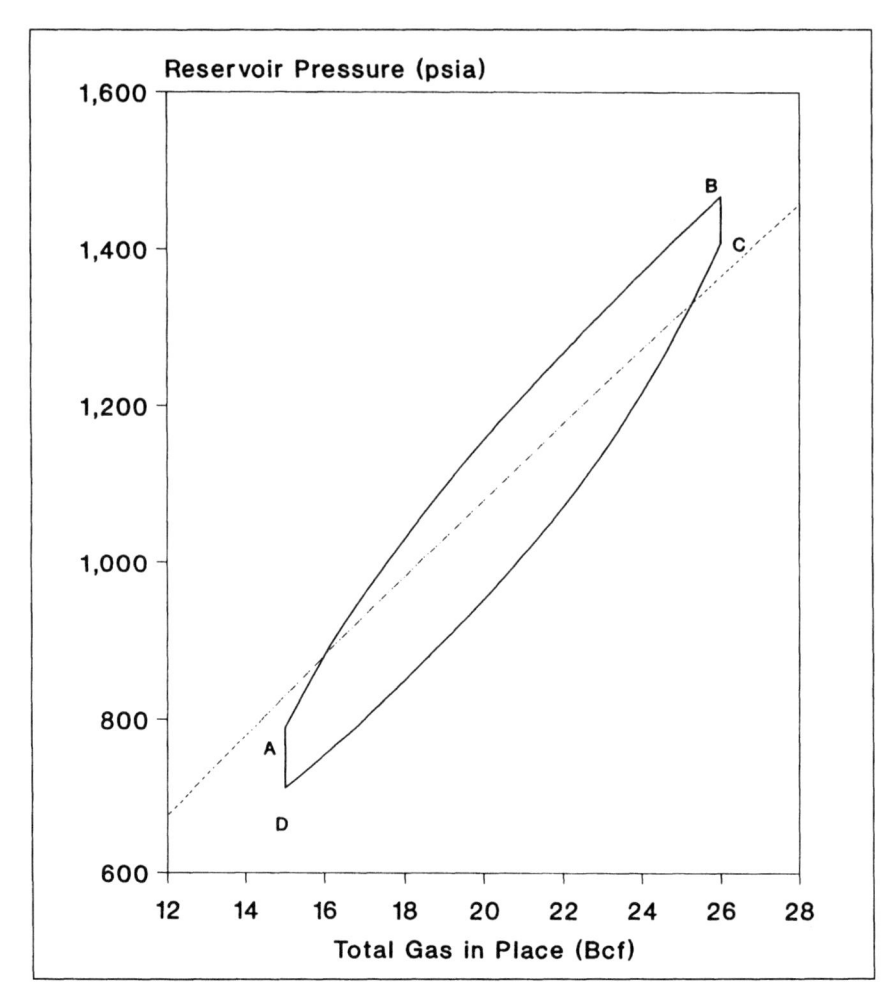

Figure 8-4. Typical injection-withdrawal cycle for a water drive reservoir.

siderably. At point C after the shut-in, the pressure is still higher than in the volumetric case. Part of this is because the pressure is not equalized in the gas portion of the reservoir. The remaining reason is because the gas has not entirely pushed back the water to an equilibrium level.

Line CD shows the withdrawal cycle. At the end of this period point D has a lower pressure than the volumetric case due to the fact that the water has not come in to fill up the pore spaces left by the gas. This causes the volume available for gas storage to be larger. The line DA is the shut-in period after the withdrawal season. During this shut-in time the pressure at the well rises to point A. The pressure at this point is still considerably lower than the pressure decline line due to the pressure transient conditions in the reservoir and due to the fact that the influx of water is not complete.

Storage Reservoir Development. Figures 8-2 and 8-4 represent reservoirs that have been developed into storage units, have reached stable and repeatable operating cycles, and are not leaking. During the development and filling of a reservoir with gas there are transition cycles. For a volumetric reservoir the development of the pressure-volume history may look like Figure 8-5. After the cushion gas is injected, a portion of the working

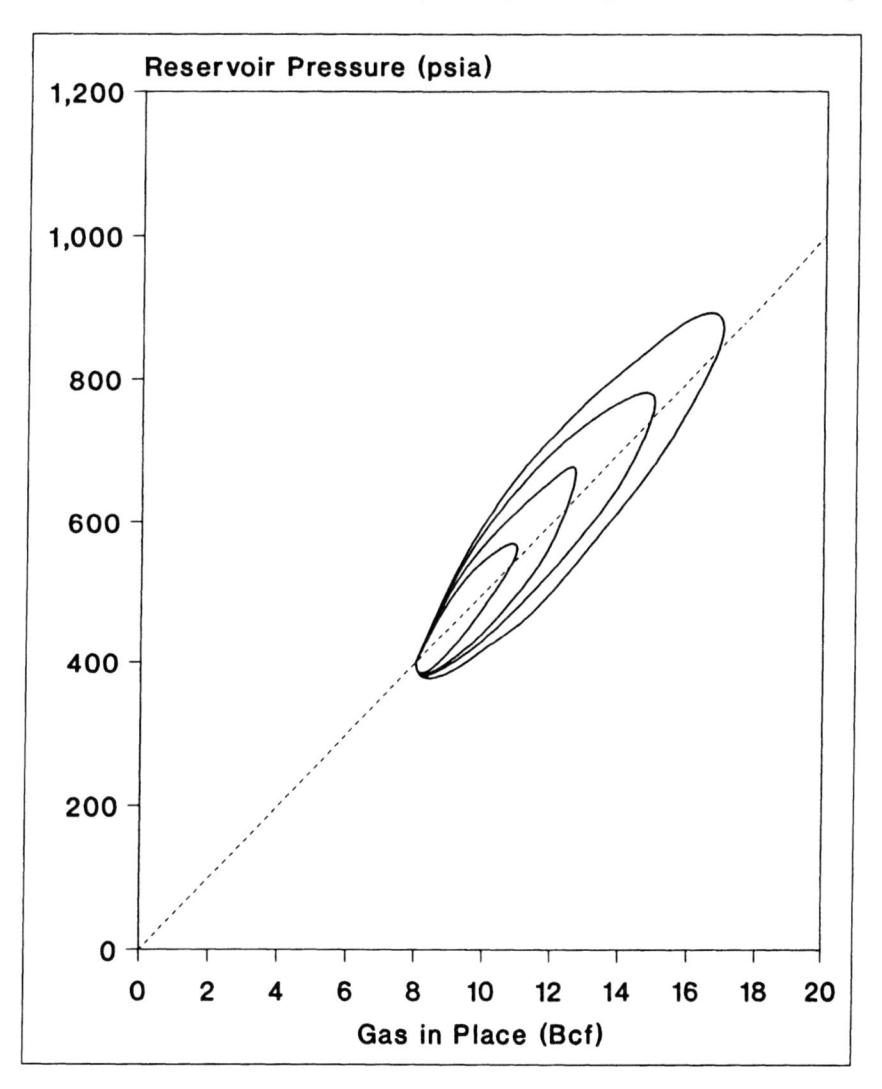

Figure 8-5. Development of a volumetric storage reservoir.

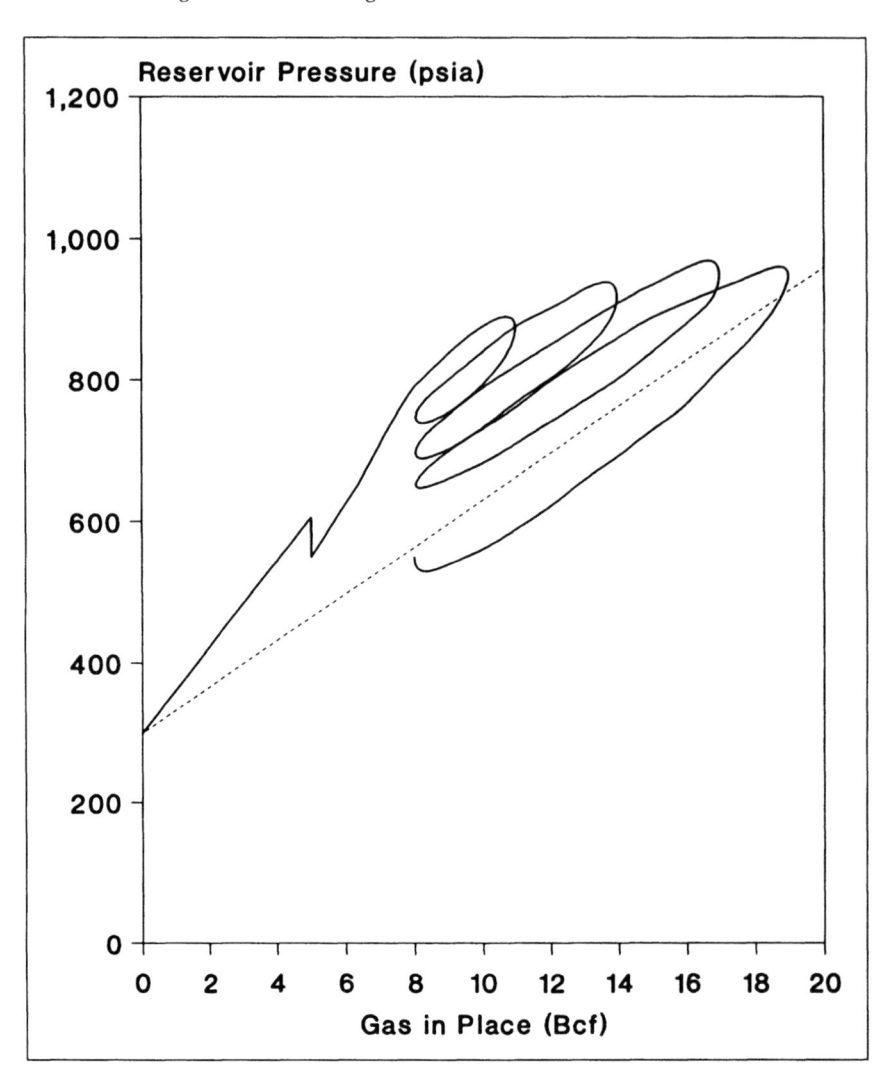

Figure 8-6. Development of a water drive reservoir.

gas is injected the first year. Following the injection season, the small amount of working gas that was injected is withdrawn. The second year the working gas that was withdrawn is reinjected plus some additional working gas is injected. This pattern may be followed for two or more years until the full complement of working gas has been injected. This schedule is usually governed by the availability of gas to inject into the reservoir. The pressure-volume cycles during this growth period are anchored around the pressure decline curve. When all of the working gas has been injected, the facility will operate similar to Figure 8-2.

When a water drive reservoir is being developed, the pressure-volume history can look quite different. In many cases, the water drive reservoir may have been either shut in or produced at relatively low pressures for long periods of time. This has allowed the water to encroach into the gas space of the reservoir. When the injection of gas begins, the gas must drive out this water. Figure 8-6 shows how a pressure-volume diagram might look in this case. The injection of cushion gas is performed at a rate such that although the pressure is sufficient to drive out the water, the friction will not allow the water to move fast enough. As a result the pressure-volume curve goes significantly above the pressure-decline curve.

Figure 8-6 shows a portion of the cushion gas being injected in one continuous operation during one season. The reservoir is then left to sit until the next injection. The amount of injection per season may be governed by the availability of gas. It may also be limited to allow the water efflux to catch up with the gas injection. At any rate, the reservoir is shut in at the end of the first injection season and the pressure decreases. Following this, the remainder of the cushion gas and a small portion of the working gas are injected during the second season. During the winter the working gas is withdrawn. The remaining cycles inject the total amount of working gas. During this time, the pressure-volume cycles move to the right of the diagram, getting closer to the pressure-decline curve as the water continues to be driven out. After several cycles the operation will resemble Figure 8-4.

Leaking Reservoirs. All of the pressure-volume diagrams discussed so far have been for reservoirs that had no leaks or inaccurate metering. Since the storage facilities operate in cycles, the same (or similar) point in a cycle can be compared for two different years. If the points are the same, it can be inferred that there is no leakage. Usually it is not possible to compare exactly the same points in the cycle due to operational differences from year to year. Because of this it is sometimes necessary to interpolate values or to make other adjustments in order to get a valid comparison.

Gas in a reservoir is stored in the pore spaces available to the hydrocarbon gas. This is called the hydrocarbon pore volume (HCPV). If the HCPV, the reservoir pressure, and the temperature are known, the amount of gas in the reservoir can be calculated. Conversely, if the gas storage volume and the reservoir temperature and pressure are known, the HCPV can be determined. In spite of the simplicity of this, there can be problems in implementing this type of calculation. If the reservoir is leaking, the amount of gas stored is uncertain. Any leak over a long period of time (several years) can cause the inferred volume in storage to be in error significantly. One method [1] to get around this is to use more timely measurements that do not cover extremely long periods of time.

The basic gas law states:

$$\frac{P_1 \times V_1}{T_1 \times Z_1} = \frac{P_2 \times V_2}{T_2 \times Z_2} \tag{8-1}$$

P = pressure
V = volume
T = temperature
Z = compressibility factor

Applying this equation to the case under consideration:

$$\frac{P_s \times Q_1}{T_s} = \frac{P_1 \times HCPV}{Z_1 \times T_R} \tag{8-2}$$

$$\frac{P_s \times Q_2}{T_s} = \frac{P_2 \times HCPV}{Z_2 \times T_R} \tag{8-3}$$

Q_1 = stored volume in reservoir at time 1 in MMScf
Q_2 = stored volume in reservoir at time 2 in MMScf
P_s = standard pressure base in psia
T_s = standard temperature base in degrees Rankine
P_1 = reservoir pressure at time 1 in psia
P_2 = reservoir pressure at time 2 in psia
T_R = reservoir temperature in degrees Rankine
Z_1 = reservoir compressibility factor at time 1
Z_2 = reservoir compressibility factor at time 2

If these two equations are solved for Q_1 and Q_2 and the resulting equations are subtracted, the result is:

$$Q_1 - Q_2 = \frac{T_S \times HCPV}{P_S \times T_R} \times \left[\left(\frac{P}{Z}\right)_1 - \left(\frac{P}{Z}\right)_2 \right] \quad (8\text{-}4)$$

Solving for HCPV:

$$HCPV = \frac{P_S \times T_R \times (Q_1 - Q_2)}{T_S \times \left[\left(\frac{P}{Z}\right)_1 - \left(\frac{P}{Z}\right)_2 \right]} \quad (8\text{-}5)$$

Equation 8-5 is not dependent on the calculated amount of gas in storage. It is dependent on the amount of gas injected or withdrawn during a relatively short period of time. This period of time may be 4 or 5 months. During this time period the rate of leakage is usually not enough to affect the results. Once the HCPV is known, the total gas in place can be calculated by the equation:

$$GIP = \frac{P_R \times HCPV \times T_S}{T_S \times T_R \times Z_R} \quad (8\text{-}6)$$

This procedure works well for a volumetric reservoir. In the case of a water drive reservoir, the HCPV changes as the gas is injected or withdrawn, and this furnishes an additional complication. One solution to this is to estimate the amount of change in the HCPV between the start and end of the injection or withdrawal period. Using the basic gas law equation and applying it to this situation:

$$Q_1 = \frac{T_S \times P_1 \times HCPV_1}{P_S \times T_R \times Z_1} \quad (8\text{-}7)$$

$$Q_2 = \frac{T_S \times P_2 \times HCPV_2}{P_S \times T_R \times Z_2} \quad (8\text{-}8)$$

$$\Delta Q = Q_1 - Q_2 = \frac{T_S}{P_S \times T_R} \times \left[\frac{P_1 \times HCPV_1}{Z_1} - \frac{P_2 \times HCPV_2}{Z_2} \right] \quad (8\text{-}9)$$

But:

$$\frac{T_S}{P_S \times T_R} = \left[\frac{Q_1 \times Z_1}{P_1 \times HCPV_1} \right] \qquad (8\text{-}10)$$

$$\Delta Q = \frac{Q_1 \times Z_1}{P_1 \times HCPV_1} \times \left[\frac{P_1 \times HCPV_1}{Z_1} - \frac{P_2 \times HCPV_2}{Z_2} \right] \qquad (8\text{-}11)$$

$$Q_1 = \frac{\Delta Q \times P_1 \times HCPV_1}{Z_1 \times \left[\dfrac{P_1 \times HCPV_1}{Z_1} - \dfrac{P_2 \times HCPV_2}{Z_2} \right]} \qquad (8\text{-}12)$$

$$Q_1 = \frac{\Delta Q \times \left(\dfrac{P}{Z}\right)_1}{\left(\dfrac{P}{Z}\right)_1 - \left(\dfrac{P}{Z}\right)_2 \times \dfrac{HCPV_2}{HCPV_1}} \qquad (8\text{-}13)$$

Tek [2] developed this same equation using a slightly different route. This method requires the estimation of the change in HCPV due to the change in water influx or efflux. The accuracy of this estimation is usually not high at best. Leak detection in reservoirs with an active water drive continues to pose a challenge to the pipeline industry, but some work is being done on this subject [3].

In stable, nonleaking reservoirs where the gas measurement is good, the pressure volume history cycle will operate within the same general region over a long period of time. If a reservoir is having gas leaked into it, or if the measurement errors result in a deficiency of inventory, the pressure-volume history will move to the left, which is illustrated by Figure 8-7. If the reservoir is leaking gas out of the reservoir, the pressure-volume history will move to the right, which is illustrated by Figure 8-8.

Example of a Large Leak. In some cases the rate of leakage is so great that the results obtained with Equation 8-5 will give erroneous results. This can be determined in a volumetric reservoir because the HCPV values calculated during injection and during withdrawal will be different. The HCPV values calculated from withdrawal data will be lower than they should be and those calculated from the injection data will be larger than they should be. Figure 8-8 illustrates a reservoir with a leak large

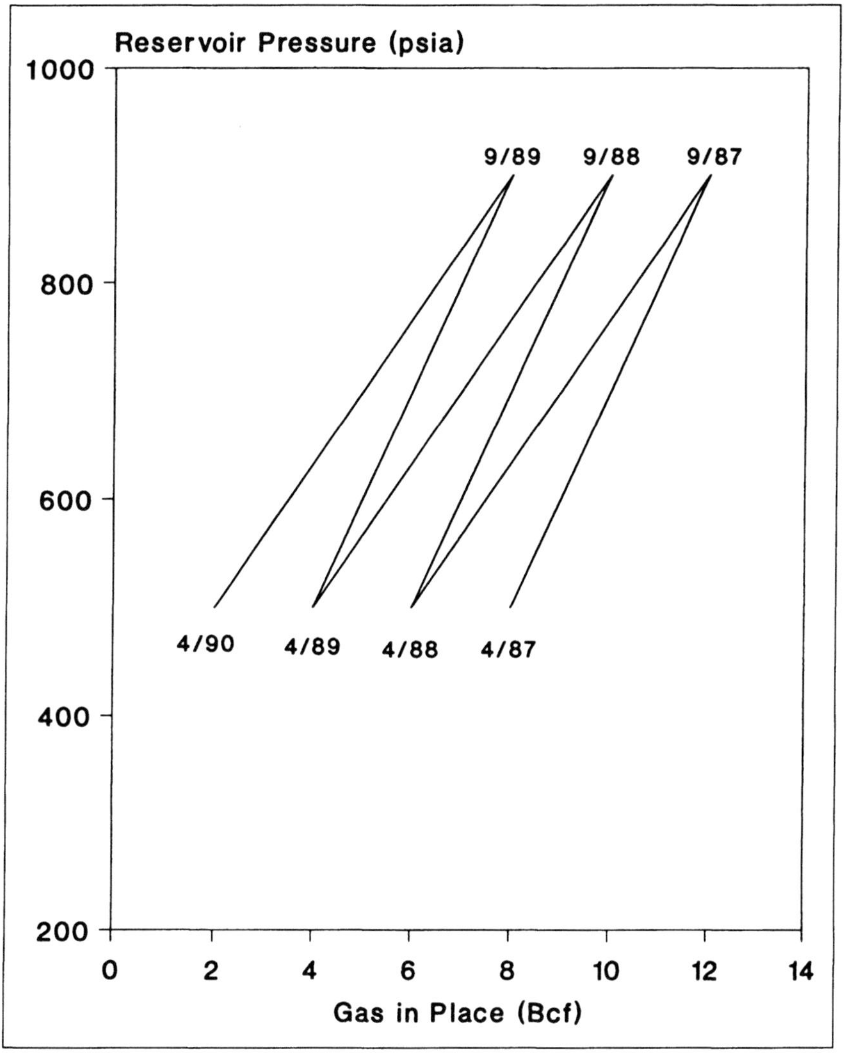
Reservoir Pressure (psia)
1000
800
600
400
200
9/89
9/88
9/87
4/90
4/89
4/88
4/87
0
2
4
6
8
10
12
14
Gas in Place (Bcf)

Figure 8-7. Example of a gas leak into a storage reservoir.

enough for this to occur. This is a small storage facility located in southern Kansas. The leak itself is not unduly large, but the reservoir is small and the leak is large in proportion.

A series of HCPV calculations have been run for the various injection and withdrawal seasons represented in Figure 8-8. The results are shown in Table 8-1. It can be seen that the values of HCPV calculated vary significantly. The last calculation covers a period of 4 years, during which time

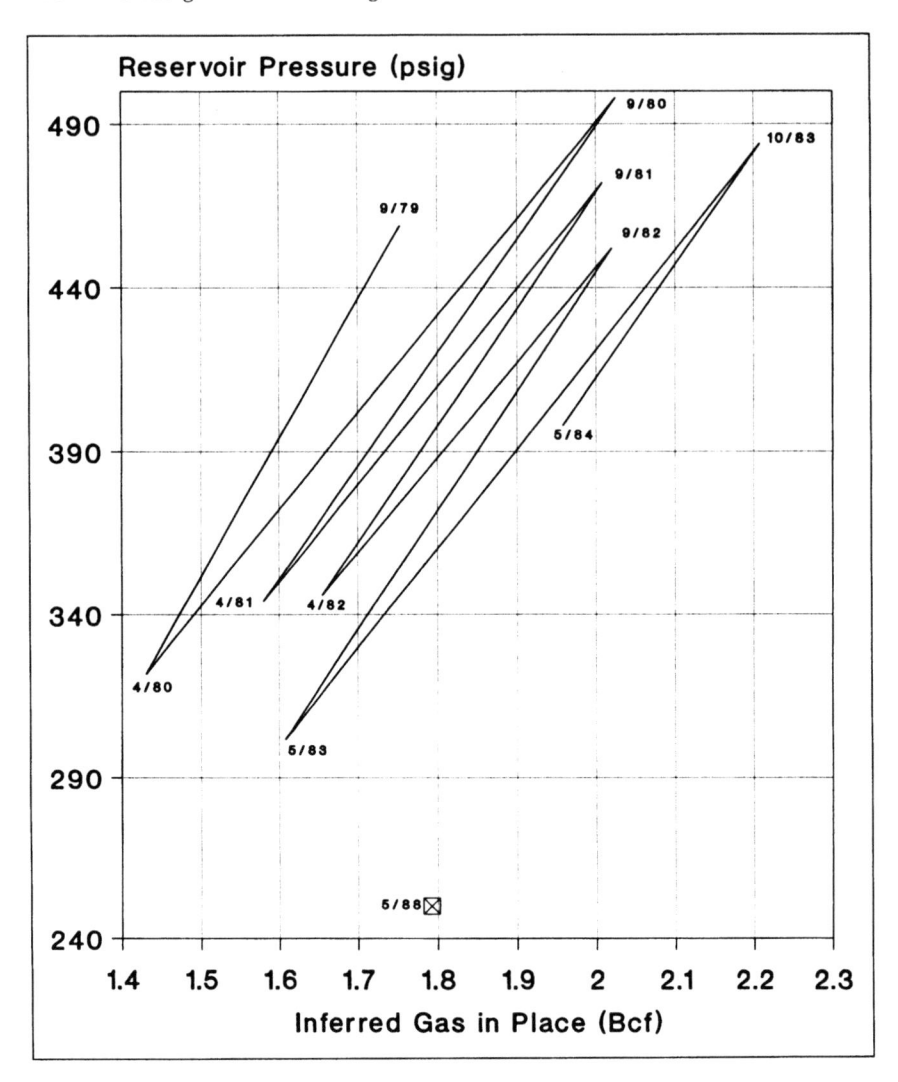

Figure 8-8. Example of a leaking volumetric reservoir.

the effect of the leak is quite large. The results from Table 8-1 are inconclusive at best. This particular case requires some further manipulation in order to give consistent results.

It is logical that the leakage rate in a reservoir will vary with pressure. Due to the dynamic operation of the reservoir, however, there is a reasonable possibility that the average pressure in the reservoir during the withdrawal cycle (November through March) is quite close to the average pressure during the injection cycle (April through October). Since this

Table 8-1
Calculation of Hydrocarbon Pore Volume Assuming No Leakage

Test Date No. 1	Test Date No. 2	Stable BHP No. 1 psig	Stable BHP No. 2 psig	Gas Injected or Withdrawn MMcfd	Calculated HCPV in MMcfd
9/79	4/80	480.3	334.9	−334.248	30.55
4/80	9/80	334.9	520.3	606.441	43.16
9/80	4/81	520.3	356.5	−452.643	36.32
4/81	9/81	356.5	493.3	436.543	42.14
9/81	4/82	493.3	359.3	−360.944	35.54
4/82	9/82	359.3	472.3	362.467	42.48
9/82	5/83	472.3	311.7	−409.693	34.08
5/83	10/83	311.7	506.4	605.717	41.31
10/83	5/84	506.4	414.0	−258.385	36.46
5/84	11/88	414.0	259.0	−154.777	13.60

concept appears reasonable, the assumption has been made that the leakage rate is constant with time. Using this assumption, a calculation of HCPV can be made for each pair of points listed in Table 8-1.

Table 8-2 shows the results of this calculation. For each pair of points the time interval between the pressure tests is shown. A leakage rate in Mcfd is assumed. This leakage rate is multiplied by the number of days between pressure tests to give the total assumed leakage between the test points. This total leakage is algebraically deducted from the gas injected or withdrawn to give the total change in gas volume. The HCPV is then calculated for each pair of points. This is a trial and error procedure that is continued until the best and most consistent set of HCPV values is obtained. The criterion for acceptance used in this case was the least value of:

$$\sum (HCPV_i - HCPV_{avg})^2$$

The best fit for this calculation was for a leakage rate of 201 Mcfd. Although this leakage rate is not unduly high in absolute terms, the leak

Table 8-2
Calculation of Hydrocarbon Pore Volume Assuming Leakage of 201 Mcfd

Test Date No. 1	Test Date No. 2	Stable BHP No. 1 psig	Stable BHP No. 2 psig	Mid–Point to Mid–Point Test Interval in Days	Inferred Change in GIP MMcfd	Calc. HCPV MMcfd
9/79	4/80	480.3	334.9	215	−377.463	34.497
4/80	9/80	334.9	520.3	158	574.683	40.903
9/80	4/81	520.3	356.5	206	−494.049	39.643
4/81	9/81	356.5	493.3	146	407.197	39.306
9/81	4/82	493.3	359.3	219	−404.963	39.876
4/82	9/82	359.3	472.3	162	329.905	38.665
9/82	5/83	472.3	311.7	219	−453.712	37.744
5/83	10/83	311.7	506.4	167	572.15	39.025
10/83	5/84	506.4	414.0	212	−300.997	42.470
5/84	11/88	414.0	259.0	1,444	−445.021	39.113
			Average of Calculated HCPV Values			39.124

is from a very small reservoir. With the small reservoir even a small leak is significant.

References

1. Flanigan, Orin. *A Definition of the Leakage in a Storage Reservoir.* Unpublished internal report. December, 1989.
2. Tek, M. R. *Underground Storage of Natural Gas.* Houston: Gulf Publishing Company, 1987.
3. Tek, M. R. *Verification of Inventory and Assurance of Deliverability in Underground Storage.* Unpublished manuscript, 1994.

CHAPTER 9

Pressure Measurements on Reservoirs

Background

All reservoirs need to have their pressure measured periodically. Producing fields need to have the pressure measured every few years in order to check the pressure decline history and to make sure that the decline rate is still on the curve. Figure 7-1 in Chapter 7 shows a typical pressure decline curve for a producing reservoir. Storage reservoirs usually have their pressure checked twice a year. This usually occurs at the end of the injection period and at the end of the withdrawal period. The pressure check on storage reservoirs is to make an inventory assessment and to determine the general health of the reservoir.

When pressure measurements are made on either a producing field or a storage field, the reservoir is usually shut in for a period of time to allow the pressure to stabilize. This time period may be 4 or 5 days for a producing well. For a storage field, however, the period may be as long as 30 days. The pressure is then measured with a dead-weight tester. After this measurement the producing well is put back in service and the storage field is returned to the "available for use" status.

The reason for shutting in the reservoirs before the pressure measurement is to allow the pressures in the reservoir to stabilize and equalize throughout the reservoir. The problem with this procedure is that the time allowed for shut-in is usually not sufficient for this pressure stabilization to occur. Figure 9-1 shows the pressure history of an actual storage field that has been shut in after the withdrawal period. During the first 45 days after shut-in there is a substantial rise in the pressure. Further, the pressure is still rising at the end of 45 days. Figure 9-2 shows the pressure history of the same reservoir when it has been shut in after the injection period. The same characteristics are present in Figure 9-2 as in Figure 9-1. In both cases the pressures are still changing after 45 days. Calculations have shown that in the cases examined, a time period of 100 to 200 days is

110 *Underground Gas Storage Facilities*

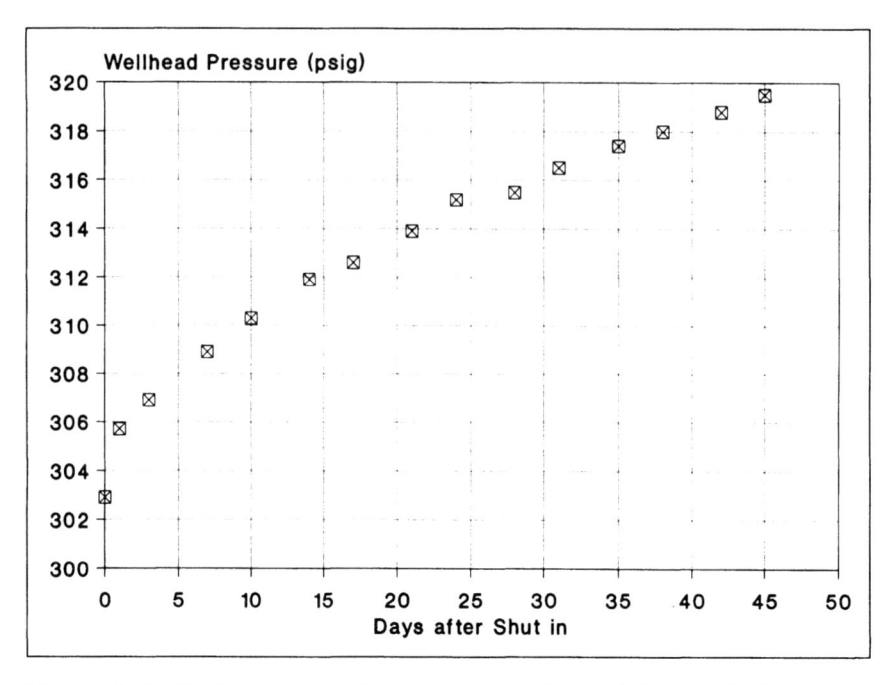

Figure 9-1. Shut-in reservoir pressure pattern at the end of a withdrawal period.

required for the complete stabilization of pressures within the field. Obviously a producing well or a storage field cannot be taken out of service and shut in for this period of time. This also means that a large portion of the pressure measurements on reservoirs are inaccurate enough to give erroneous results in the calculations that use these pressure measurements.

Theoretical Development

A method is needed to take the pressure measurements made on a reservoir over a reasonable shut-in period and arrive at a stabilized pressure that can be considered accurate. In order to pursue this, Flanigan [1] has assumed a permeable reservoir with an active water drive. As gas is injected, the water is pushed out of the reservoir; this is illustrated in Figure 9-3. When injection stops, the pressure continues to push the water down until an equilibrium pressure is reached. The rate of water movement is proportional to the difference between the current pressure and the equilibrium pressure.

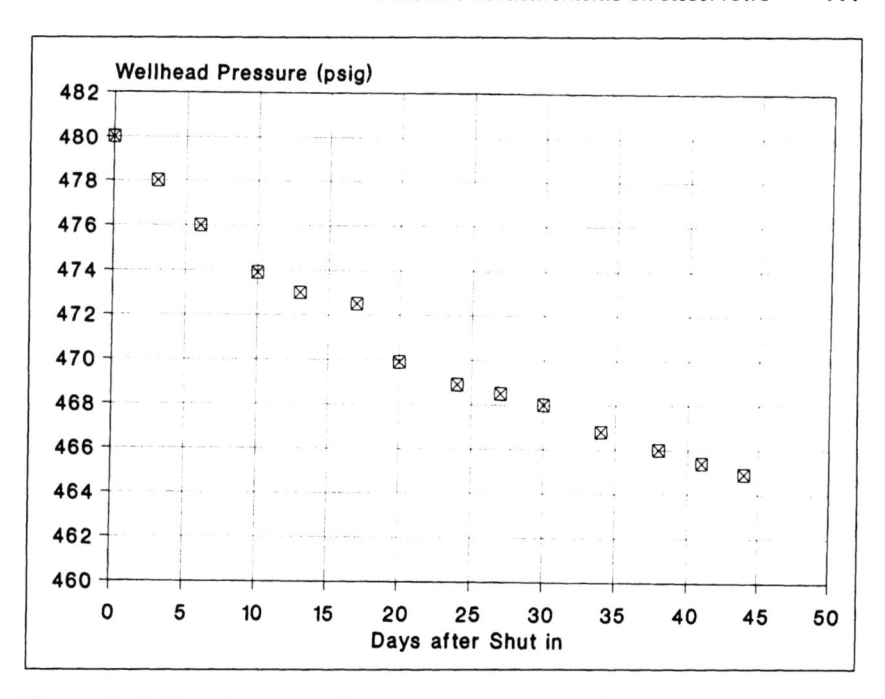

Figure 9-2. Shut-in reservoir pressure pattern at the end of an injection period.

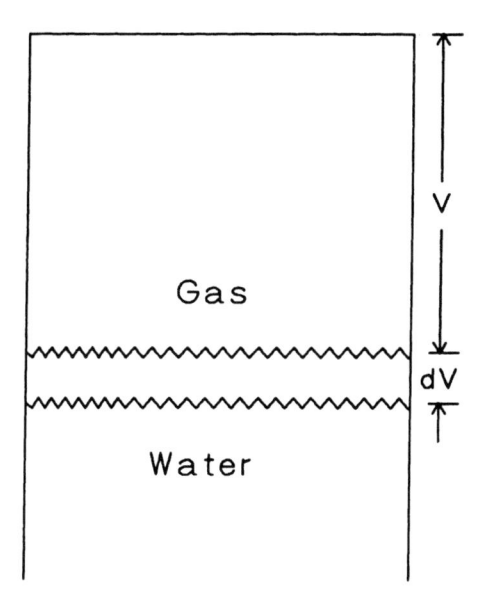

Figure 9-3. Schematic diagram of a water drive reservoir.

$$\frac{dV}{dt} = e \times (P - P_e) \tag{9-1}$$

V = reservoir volume exposed to gas
t = time
P = current reservoir pressure
P_e = reservoir pressure at equilibrium
e = proportionality constant

Figure 9-3 and Equation 9-1 apply to a water drive reservoir, regardless of the mechanism of the water drive. If the drive is due to hydrostatic head or if it is due to the compressibility of water, the equation still applies. Consider that the analogy behind the equation is broadened to include the case where gas pushes back an invisible barrier in order to expand to areas being denied to it by the time lag of friction to the flow of gas. When the gas has fully expanded into the full area of the reservoir, the pressure will reach the equilibrium pressure P_E. With this analogy, the equation and the reasoning will apply to a volumetric reservoir as well as to a water drive reservoir.

The gas law states:

$$Q = V \times \frac{P \times T_S}{P_S \times T \times Z} \tag{9-2}$$

Q = total gas in place in reservoir at standard conditions
P_s = standard pressure, usually 14.7 psia
T_s = standard temperature
T = reservoir temperature
P = reservoir pressure
Z = compressibility factor

The compressibility factor may be represented by a simplified equation:

$$Z = c + d \times P \tag{9-3}$$

c = constant
d = constant

Substituting for Z:

$$Q = \frac{P \times V \times T_S}{P_S \times T \times (c + d \times P)} \tag{9-4}$$

$$V = \frac{P_S \times T \times Q}{T_S} \times \left(\frac{c}{P} + d\right) \tag{9-5}$$

Differentiating:

$$dV = \frac{P_S \times T \times Q \times c}{T_S \times P^2} \times dP \tag{9-6}$$

$$dV = e \times (P - P_e) dt = \frac{P_S \times T \times Q \times c}{T_S} \times \frac{dP}{P^2} \tag{9-7}$$

$$\int_{P_1}^{P} \frac{dP}{P^2 \times (P - P_e)} = \frac{-eT_S}{P_S \times T \times Q \times c} \int_{t_1}^{t} dt \tag{9-8}$$

Integrating:

$$\left[\frac{1}{P \times P_e} + \frac{1}{P_e^2} \times \ln \frac{P - P_e}{P}\right]_{P_1}^{P} = \frac{-eT_S}{P_S \times T \times Q \times c} \times [t]_{t_1}^{t} \tag{9-9}$$

Substituting the values of the integration limits:

$$\left(\frac{1}{P \times P_e} + \frac{1}{P_e^2} \times \ln \frac{P - P_e}{P}\right) - \left(\frac{1}{P_1 \times P_e} + \frac{1}{P_e^2} \times \ln \frac{P_1 - P_e}{P_1}\right)$$
$$= \frac{-e \times T_S \times (t - t_1)}{P_S \times T \times Q \times c} \tag{9-10}$$

$$(t - t_1) = \frac{P_S \times T \times Q \times c}{e \times T_S}$$
$$\left[\frac{1}{P_1 \times P_e} - \frac{1}{P \times P_e} - \frac{1}{P_e^2} \times \ln \frac{(P - P_e) \times P_1}{(P_1 - P_e) \times P}\right] \tag{9-11}$$

Equation 9-11 relates time to pressure from some pressure P_1 and time t_1 to the equilibrium pressure P_e. This equation contains the three unknowns c, e, and P_e. While these unknowns could be determined by substituting three sets of values into Equation 9-11, there is a simpler way. The equation may be rearranged to the following:

$$\frac{e \times T_S}{P_S \times T \times Q \times c} = \frac{\left[\dfrac{1}{P_1 \times P_e} - \dfrac{1}{P \times P_e} - \dfrac{1}{P_e^2} \times \ln\dfrac{(P - P_e) \times P_1}{(P_1 - P_e) \times P}\right]}{(t - t_1)} \quad (9\text{-}12)$$

The left-hand term of Equation 9-12 contains all of the unknowns except P_e, which is being solved for. If the right-hand portion of Equation 9-12 is substituted with one P-t point and then substituted again with another P-t point, and the two right-hand sides are set equal to each other, the following results:

$$\frac{\left[\dfrac{1}{P_1 \times P_e} - \dfrac{1}{P(1) \times P_e} - \dfrac{1}{P_e^2} \times \ln\dfrac{(P(1) - P_e) \times P_1}{(P_1 - P_e) \times P(1)}\right]}{t(1) - t_1}$$

$$= \frac{\left[\dfrac{1}{P_1 \times P_e} - \dfrac{1}{P(2) \times P_e} - \dfrac{1}{P_e^2} \times \ln\dfrac{(P(2) - P_e) \times P_1}{(P_1 - P_e) \times P(2)}\right]}{t(2) - t_1} \quad (9\text{-}13)$$

This equation contains only the unknown of interest, P_e.

Example

In order to solve for this, a value of P_e is assumed and substituted into the left-hand side of equation 9-13 and also into the right-hand side, and each side is evaluated. The right-hand side is then subtracted from the left-hand side and the result is examined. If the correct value of P_e was assumed, the resulting difference will be 0. In practice, values of P_e are chosen until the difference becomes sufficiently small. Using the data from Figure 9-2, pressure-time points were selected and substituted into Equation 9-13. Table 9-1 shows this example. By trial and error the equilibrium pressure is found to be 462.02 psig.

From the values determined in the example shown in Table 9-1 the pressure-time history curve can be calculated and plotted. Equation 9-12

Pressure Measurements on Reservoirs 115

Table 9-1
Example of Solving for Equilibrium Pressure

Pressure Data Selected:

	P psig	P psia
P1	480	494.7
P(1)	474	488.7
P(2)	466	480.7

Time Data Selected:

	Time In Days
t1	0
t(1)	10
t(2)	38

Trial and Error Calculation for Pe:

Assumed Pe In psia	Value of e x Ts / Ps x T x Q x c		Balance Criteria
	Calculation With P-t Point #1	Calculation With P-t Point #2	
476	0.000000160169	0.000000153803	6.3661E-09
477	0.000000171378	0.000000174465	-3.0868E-09
476.7	0.000000167852	0.000000167605	2.4730E-10
476.8	0.000000169011	0.000000169819	-8.0861E-10
476.72	0.000000168082	0.000000168042	4.0299E-11
476.73	0.000000168198	0.000000168262	-6.3974E-11

Pe = 476.72 psia

Pe = 462.02 psig

e x Ts / Ps x T x Q x c = 0.0000001681

Table 9-2
Example of Calculating Pressure-Time Curve

Pressure psig	Pressure psia	Time In Days	
480	494.70	0.00	
477	491.70	4.46	
474	488.70	9.99	
472	486.70	14.56	
471	485.70	17.22	
470	484.70	20.20	
469	483.70	23.59	
468	482.70	27.53	
467	481.70	32.21	
466	480.70	37.96	
465	479.70	45.43	
464	478.70	56.01	
463	477.70	74.30	
462.5	477.20	92.92	
462.4	477.10	99.02	
462.3	477.00	107.00	
462.1	476.80	139.75	
462.08	476.78	147.27	
462.03	476.73	194.14	
462.02	476.72	218.11	Stable
462.01	476.71	ERR	

is used for this. Values of pressure are substituted into the equation and the equation is solved for time t. Table 9-2 and Figure 9-4 show the result of this calculation. The stabilized pressure of 462.02 psig is reached after 218 days. It should be noted that if the assumed value of the stabilized pressure in the trial-and-error calculation overshoots the true value, an error message results. This is because the overshoot results in trying to take the logarithm of a negative number. The last entry in Table 9-2 illustrates this.

Normally the maximum time a storage facility can be shut down to measure pressure is about 30 days. From Table 9-2 it can be seen that the reservoir changes from 480 psig at initial shut-in to 467 psig after 32 days.

Pressure Measurements on Reservoirs **117**

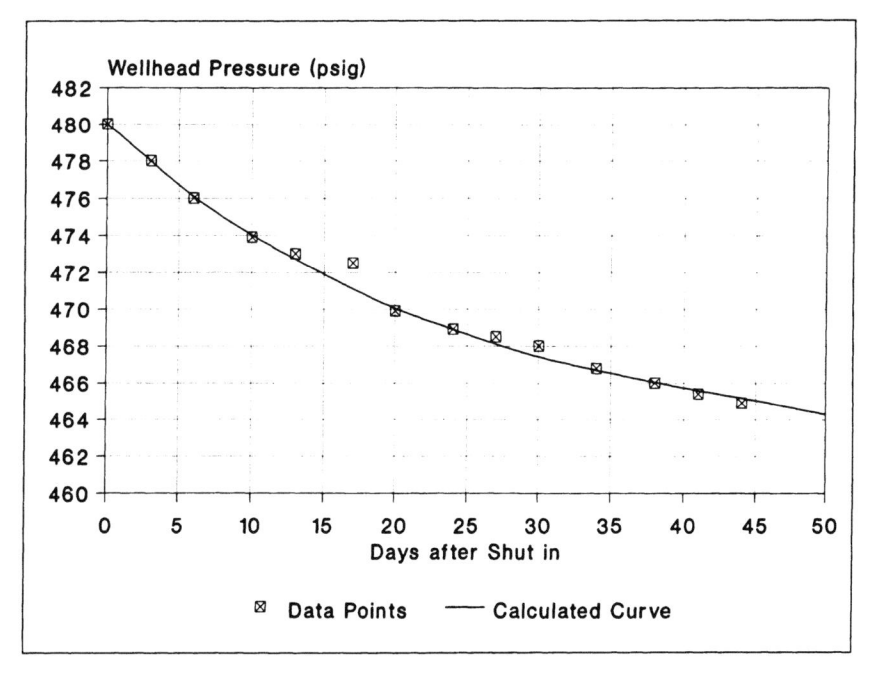

Figure 9-4. Shut-in pressure history correlation for the end of an injection period.

This change of 13 pounds pressure represents a considerable amount of equalization. At complete stabilization, however, the pressure is 462 psig. This is still 5 pounds lower than the pressure after 32 days shut-in. This lack of stabilization is enough to give significant error to conclusions that may be derived from calculations using the 32-day number. It is suggested that the calculated stabilized pressure be used in all reservoir shut-in tests, both storage fields and producing fields.

References

1. Flanigan, Orin. *A Definition of Leakage in a Storage Reservoir.* Unpublished internal report. December, 1989.

CHAPTER 10

Metering

There are two main reasons why the actual inventory of gas in an underground storage facility does not match the book or accounting quantity of inventory. One reason is the leakage of gas into or out of the reservoir. The second reason is meter error or discrepancy. Meter accuracy is particularly important for a storage facility. Anytime two meters are placed in series measuring the same stream of gas, there will be a difference between the two readings. This is the nature of metering. There is no such thing as a perfectly accurate meter. The question becomes not whether there is error, but how much the error is. A storage facility is a classic case of the two meters in series. The injection meters measure the gas going into storage. During withdrawal, a second set of meters measures this same gas as it is being withdrawn. Any error in either of the meters shows up as an error in inventory. The effect of this is cumulative with time. Even though the difference between the two sets of meters may be small, this small difference accumulates over several years and shows up as a discrepancy between booked and actual inventory. Although this error cannot be completely eliminated, there are measures that can help to lessen it.

Types of Meters

There are many types of meters in service in the gas industry today. These include the following:

Orifice meters	Sonic nozzles
Positive displacement meters	Ultrasonic meters
Flow nozzles	Turbine meters
Vortex shedding meters	Doppler meters
Coriolis meters	Insertion turbines
Pitot tubes	Elbow meters

Of all these meters, the gas industry uses primarily the orifice meter and the turbine meter. The orifice meter is used in the vast majority of storage facility installations. Because of this, the thrust of this chapter will be devoted to orifice metering and its peculiarities.

Sources of Meter Error

When the subject of error in metering is addressed, in general, there are about three areas where errors can occur. The first is in the installation of the meter. The American Gas Association [1] has published guidelines for the proper installation of orifice meters. These guidelines are used extensively in the gas industry. These are minimum standards, and when the metering installation conforms to these minimum standards the installation is considered to be acceptable for transfer accuracy. The second area for meter error consists of malfunctions in the meter installation. Examples of this are an orifice plate being installed backwards, straightening vanes breaking loose and getting too close to the orifice plate, liquid or solids accumulating in the meter run and disturbing the flow pattern of the gas approaching the orifice plate, and similar malfunctions.

The third area of metering errors involves the manner in which the orifice meter is operated. The basic equation for the metering of gas through an orifice is:

$$Q = C \times \sqrt{P \times h} \qquad (10\text{-}1)$$

Q = flow rate of the gas in MMScf per day
P = pressure downstream of the orifice in psia
h = orifice differential in inches of water
C = orifice coefficient

Because of the nonlinear relationship between differential and flow rate, the magnitude of the differential value can amplify errors. In most meter charts there is some degree of error in the placement of the pen on the chart lines. This error may be small, but it exists. Figure 10-1 shows the relationship between this pen placement error and flow rate error for various values of meter differential. The x-axis is the meter differential and the y-axis is the meter error. The parameters are the errors in pen placement on the chart. A pen placement error of 0.10 inches of water on a 100-inch differential chart is a very small amount. Yet at low differentials this small error in pen placement can yield a significant meter error. For this reason it is important to keep the meter differential as high as practical. This can be achieved by using an orifice with a smaller bore.

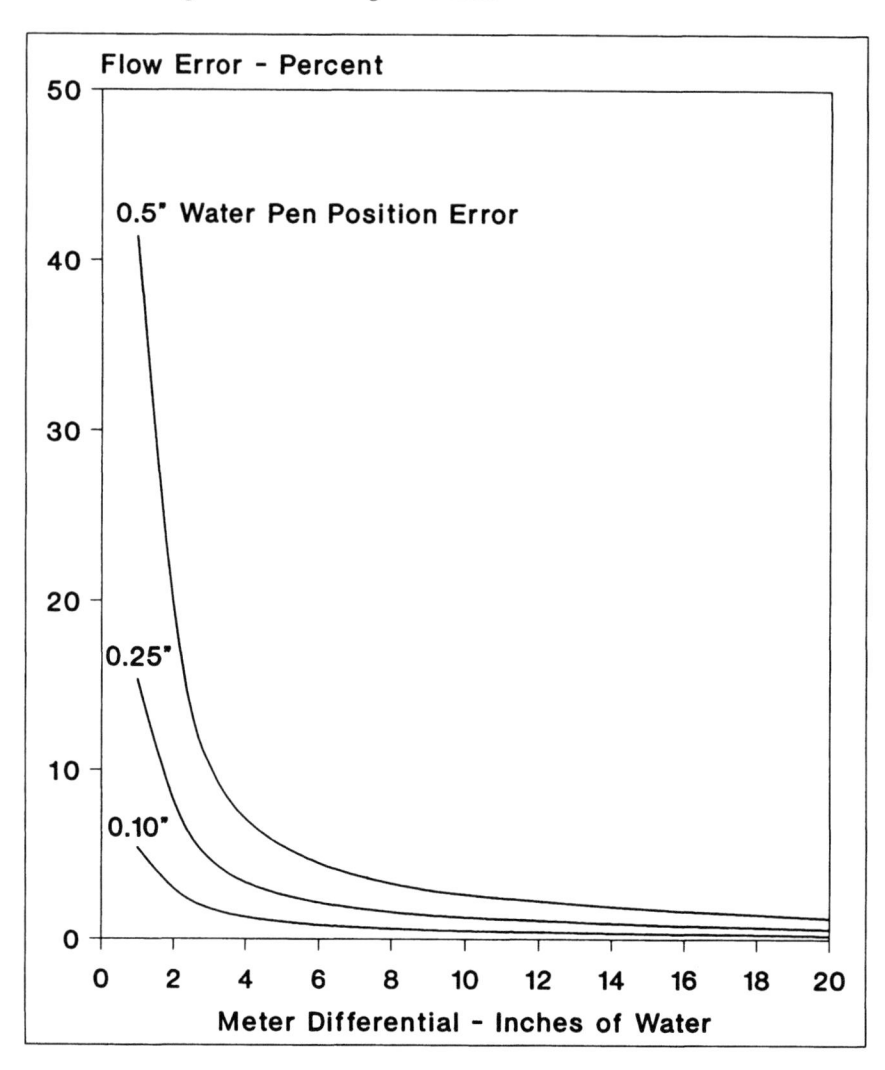

Figure 10-1. Effect of meter differential on the magnification of pen position error.

There is usually a fear on the part of field personnel to put in an orifice plate with a small bore because of the possibility of incurring a differential greater than 100 inches at times. If this occurs, the meter cannot record anything over 100 inches, and that portion of the measurement is lost. While this is a valid concern, it can be tempered by seasonal adjustment of the orifice size.

The same problem can be experienced with a low static pressure. This is not normally a problem because the static pressure is usually sufficiently high. In addition, the static pressure does not vary as much as the differential pressure.

Current Research

Recently there has been a great deal of research performed on the effect of upstream disturbances on the accuracy of orifice meters. The National Institute of Standards and Technology (formerly the National Bureau of Standards) has used laser technology to investigate turbulence upstream of orifice plates. This work has shown that with two elbows out of plane, there is swirl detectable as far as 155 pipe diameters downstream of the disturbance. This type of information infers that it is not practical to rely on lengths of pipe to allow any disturbance in the gas stream to dissipate before the orifice plate is reached. It appears that it will be necessary to use flow conditioners to restore the gas stream flow profile to an acceptable shape before it reaches the orifice plate. Considerable work is being done in this area. New designs for flow conditioners are being tested, and the results are optimistic.

Pulsating Flow

Background. Many of the facets of metering errors have been extensively discussed in the literature [2]. There is one aspect, however, that has not received a great deal of attention. This is the effect of transients, particularly pulsations, on orifice meter accuracy. The subject of pulsation came to prominence around 1950 when individual pipeline companies discovered there were symptoms of pulsations at their reciprocating compressor stations. Hold-down straps broke, vibration was excessive, and piping failed. An investigation revealed that other transmission companies were having similar problems. In order to attack the problem, 13 gas companies and compressor vendors formed a group called the Pulsation Research Council under the administration of the Southern Gas Association. The group put up some money and retained Southwest Research Institute (SwRI) in San Antonio to study the problem.

The effort at SwRI was highly successful. The theory of pulsating flow generated by reciprocating compressors was formulated, and an analog was developed that could simulate the reciprocating compressors and their associated piping. This allowed new compressor designs to be tried out on the analog before they were installed in the field. Similarly, modifications to existing stations could be tested in order to determine the effect they would have.

The Pulsation Research Council patented this technology and charged a royalty to companies to use the analog and associated technology. This royalty was used to fund additional research work in pulsation, vibration, and other related topics. Over its life the project has generated over $20,000,000, which has been plowed back into additional research. In order to reflect the wider interest, the name of the organization was changed to the Pipeline and Compressor Research Council (PCRC).

One of the areas of research was the metering of pulsating flow. In the mid-1960s the PCRC investigated metering water under pulsating flow conditions. Although good correlation was established for liquids, no similar correlation could be found for gases, and the research work was put in abeyance.

In 1975 a new attempt was made to solve the problem of the metering of gases under pulsating flow, and this time the effort was successful. A correlation, called the square root error theory, was developed and validated. This theory states:

1. The error in orifice meters at the orifice taps could largely be explained by the square root error. Although other phenomena could exist, these were not normally encountered under most operating conditions.
2. Square root error always causes an orifice meter to read high.

Field Tests. When this information became available, Arkla Pipeline Group, a Shreveport, Louisiana, based transmission company, decided that it should examine it in relation to its field installations. In order to accomplish this, a pulsation filter was designed to be placed between the field compressor and the meter run at a wellhead installation where gas for the pipeline was being purchased from a producer. This pulsation filter was tested on the PCRC analog at San Antonio, and five of the units were built.

Figure 10-2 shows the general elements of a pulsation filter. It consists of two volume bottles of equal diameters and lengths connected at their mid points by a choke tube whose length is the same as the bottle length. Gas enters at the quarter length point of one bottle and exits at the half length point of the other bottle. Figure 10-3 shows how the theoretical elements can be fabricated into a single vessel.

Metering 123

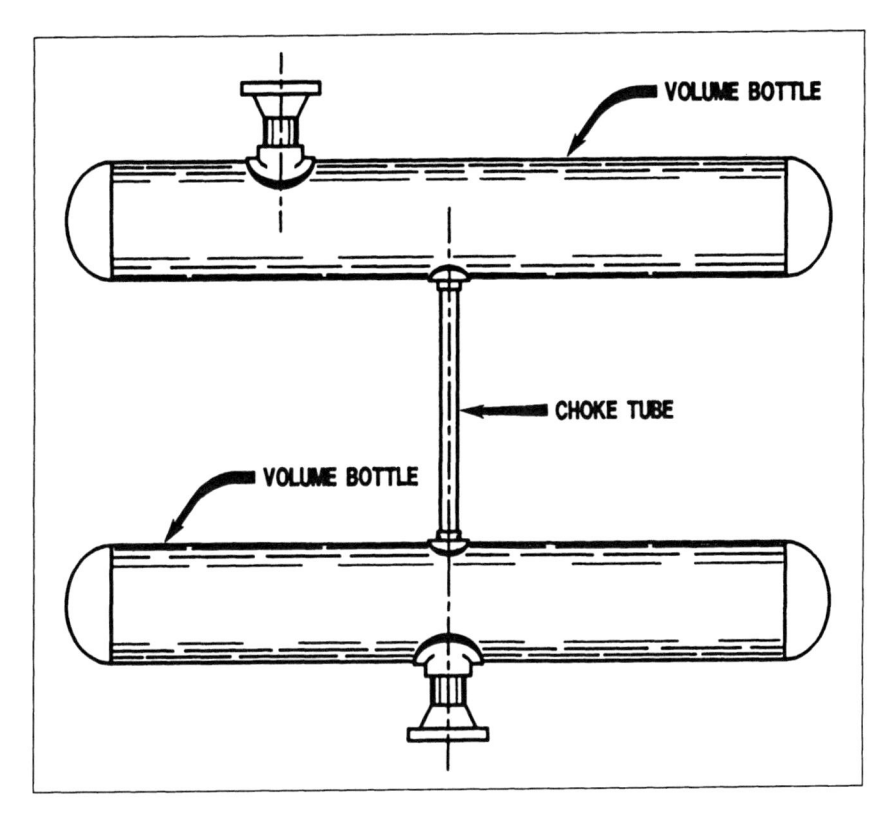

Figure 10-2. Basic elements of a pulsation filter.

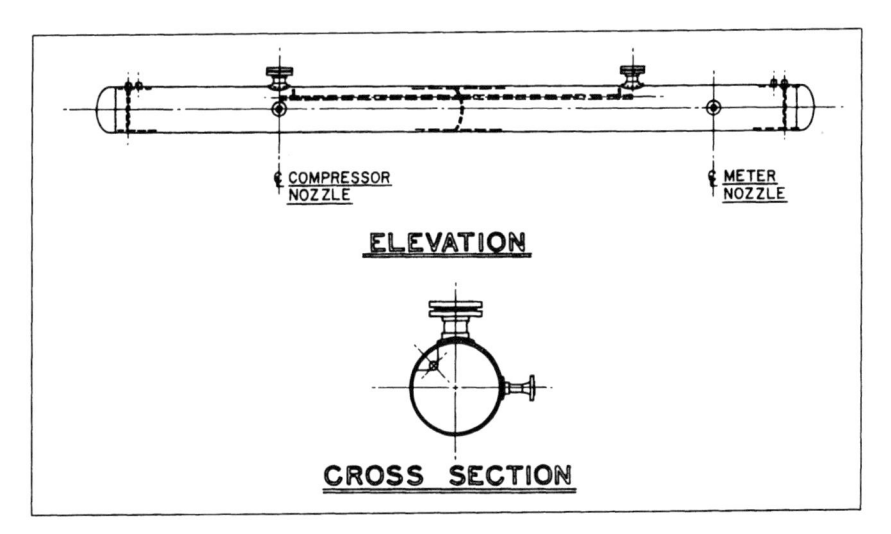

Figure 10-3. Standard pulsation filter.

124 Underground Gas Storage Facilities

Five of these pulsation filters were fabricated and installed on five well sites where the orifice meters were suspected of having pulsating flow. The filters were installed with block and bypass valves so that they could be valved in or out of service easily. Figure 10-4 is a schematic diagram showing a typical installation.

The tests consisted of putting a "fast" clock on the orifice meter chart so that the chart made one revolution per hour. With the compressor running and conditions stabilized, the suction pressure, discharge pressure, and rpm were recorded with the pulsation filter out of service and the chart was allowed to run for several minutes. The filter was then placed in service and the same measurements were taken. The filter had been designed for negligible pressure drop. This procedure was carried out at several compressor speeds.

The meter pressures, temperatures, differentials, and other information were used to make an AGA #3 volume calculation for each compressor speed with the pulsation filter in and out of service. The results for the five wells showed a minimum error of 12 percent and a maximum error of 56 percent. Figure 10-5 shows the test chart from one of the wells tested, and

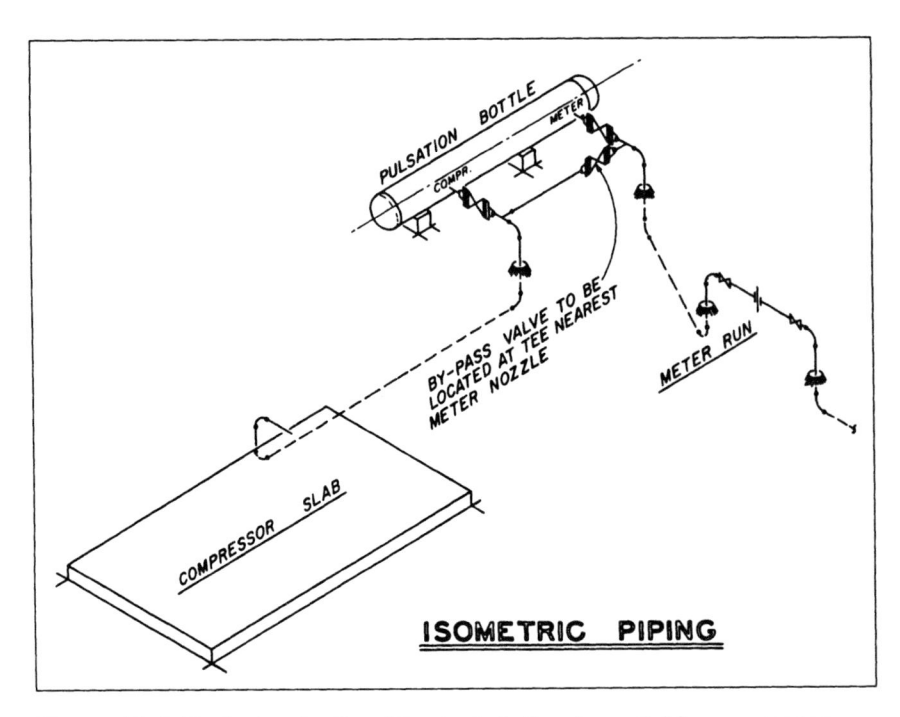

Figure 10-4. Typical pulsation filter installation for a field compressor.

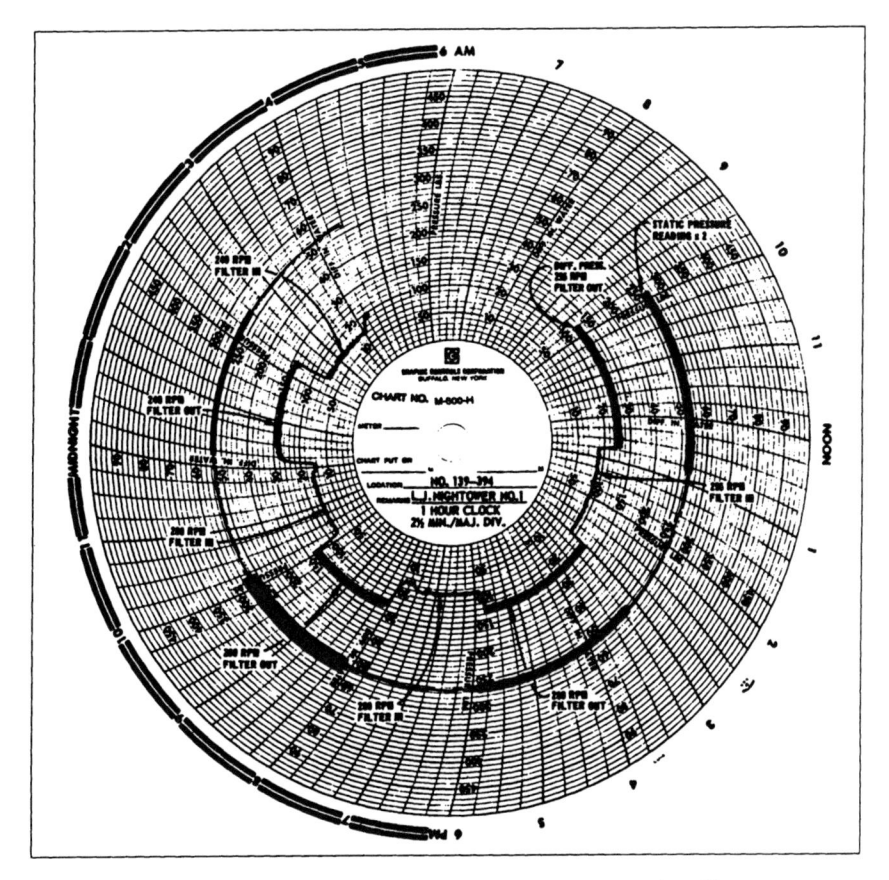

Figure 10-5. Meter chart showing the effect of a pulsation filter.

Table 10-1 lists the results of the AGA #3 volume calculations for each segment of that well test. In perspective, these five wells were chosen because they were suspected of having high meter error.

Figure 10-6 shows a high-speed recording of the differential pressure across the orifice taps during one of the tests with the pulsation filter out of service. It can be seen that the primary pulsation frequency is about 5 cycles per second or 300 cycles per minute. This corresponds to the running speed of the Ajax compressor. The peak-to-peak amplitude of the pulsations is about 125 inches of water. This is severe pulsation.

Square Root Error. Square root error is defined as the difference between the average of the square roots and the square root of the average. This error is strictly mathematical, not a matter of physics. In order

Table 10-1
Results of Pulsation Filter Test

Compressor Speed in RPM	Pulsation Filter Status	Meter Differential Pressure in Inches	Indicated Flow Rate in MMcfd	Meter Error
295	out	25.0	1.141	17.9%
295	in	18.0	0.968	
280	out	25.0	1.141	19.5%
280	in	17.0	0.955	
260	out	26.0	1.164	27.5%
260	in	16.0	0.913	
240	out	28.5	1.218	37.8%
240	in	15.0	0.884	

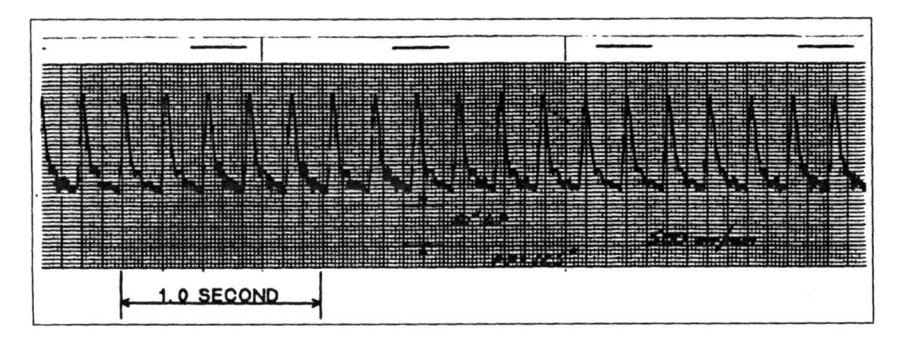

Figure 10-6. Differential pressure recording at an orifice with the pulsation filter out of service.

to illustrate square root error, Figure 10-7 was created. The figure represents an orifice meter that has an average differential of 45 inches of water and an impressed pulsation of plus and minus 36 inches (72 inches peak to peak). A square wave is used to simplify the mathematics of the example. The frequency of the impressed pulsation is 5 cycles per second, representing an Ajax compressor near its minimum speed.

During time period 0.1 to 0.2 the differential is 81 inches of water. During time period 0.2 to 0.3 the differential falls abruptly to 9 inches of water. Since the orifice meter chart pen cannot follow this rapid change in

DIFFERENTIAL IN INCHES

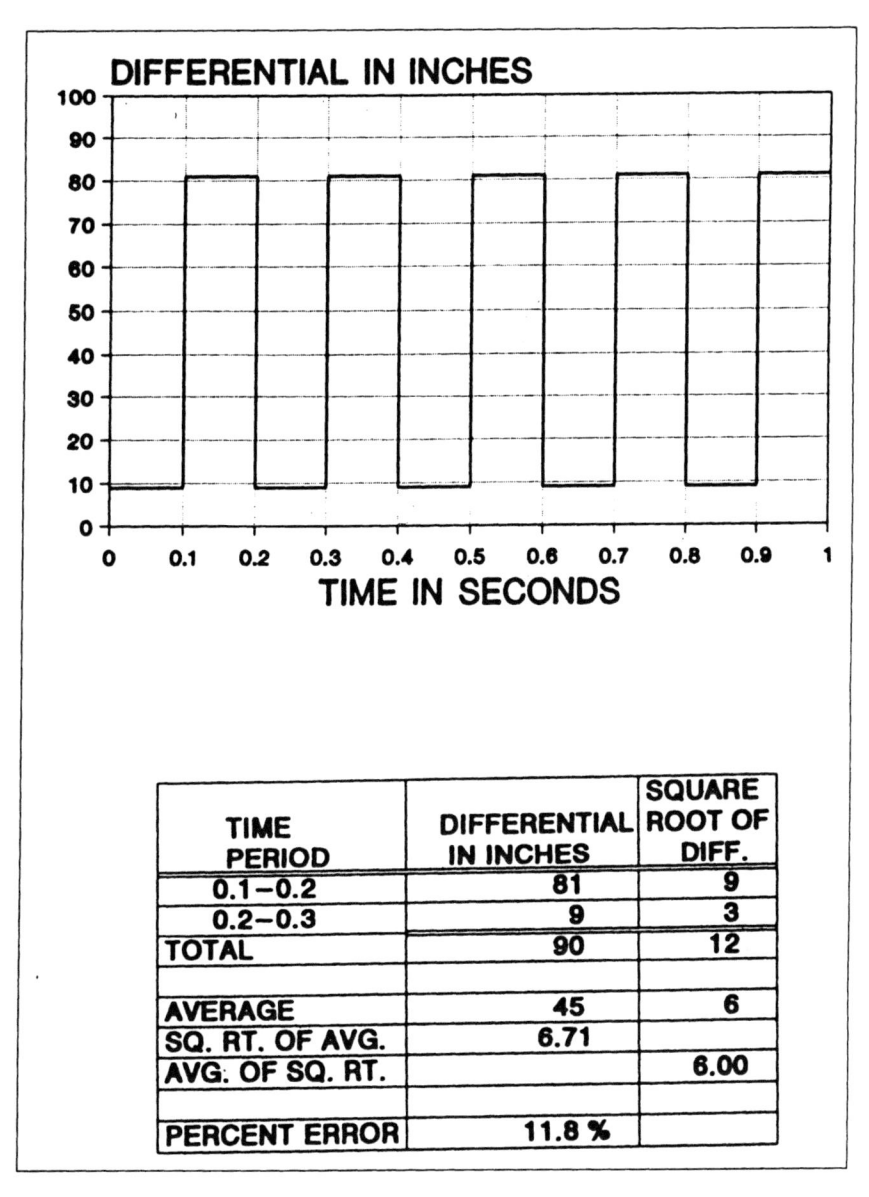

TIME PERIOD	DIFFERENTIAL IN INCHES	SQUARE ROOT OF DIFF.
0.1–0.2	81	9
0.2–0.3	9	3
TOTAL	90	12
AVERAGE	45	6
SQ. RT. OF AVG.	6.71	
AVG. OF SQ. RT.		6.00
PERCENT ERROR	11.8 %	

Figure 10-7. Example of square root error.

differential, the pen records the average differential of 45 inches. When the meter chart is integrated this average differential has its square root extracted and the resulting square root of the average is 6.71 units proportional to flow.

Table 10-2
Time for Full-Scale Pen Travel on an Orifice Meter Chart-Barton 202E Meter

Direction of Pen Travel	Percent Dampner Open	Pen Travel Time in Seconds	Direction of Pen Travel	Percent Dampner Open	Pen Travel Time in Seconds
0–100	100	6.9	100–0	100	7.8
0–100	100	7.0	100–0	100	6.1
0–100	100	7.9	100–0	100	8.0
	Average	7.3		Average	7.3
0–100	50	10.0	100–0	50	13.5
0–100	50	9.9	100–0	50	14.0
0–100	50	10.0	100–0	50	13.2
	Average	10.0		Average	13.6
0–100	0	57.9	100–0	0	63.9
0–100	0	48.9	100–0	0	64.5
0–100	0	38.5	100–0	0	63.5
	Average	48.4		Average	64.0

The true flow, however is proportional to the instantaneous square root of the differential. The square roots of 81 and 9 are 9 and 3 respectively. The resulting average of the square roots is 6.00 units proportional to flow. The difference between the average of the square roots and the square root of the average gives an error in flow of 11.8 percent.

Orifice Meter Frequency Response. In order to verify the slow response of an orifice meter chart pen, a test was run on a chart meter in the laboratory. A Barton 202E meter with a differential scale of 100 inches of water was chosen and the damping valve was set fully open for the fastest pen response. A pressure of 100 inches of water was suddenly imposed on the differential cell and the time was recorded for the pen to travel full scale. The differential pressure was then suddenly reduced to 0, and the time was recorded for the pen to come to 0. The results are shown in Table 10-2. Approximately 15 seconds are required for the pen to travel from 0 to full scale and back to 0. When the pulsation frequency is 5 cycles per second, 75 cycles of pulsation would have occurred while the chart pen was trying to measure 1 cycle.

Gauge Line Error. The gauge lines between the orifice taps and the chart meter can and do amplify the square root error. The example illustrated by Figure 10-7 shows that a peak-to-peak pulsation of 72 inches of water (2.6 psi) produced a meter error of 11.8 percent. Using a similar relationship, the pulsation filter test that showed 56 percent error would have had to have pulsation magnitudes that were physically damaging to the piping. Since this was not the case, the explanation must lie in the effect of the gauge lines.

The gauge lines can amplify the flow measurement error in two ways. First, the gauge lines can tune up and become resonant to one of the pulsation frequencies. This increases the amplitude of the pulsations and results in a higher square root error at the meter chart than was present at the orifice. Secondly, the gauge lines can rectify the differential signal. This occurs when the gauge lines have enlargements and contractions and these are not identical in both lines. An example of how this can occur is shown in Figure 10-8. For an enlargement (or contraction) in the pipe, the

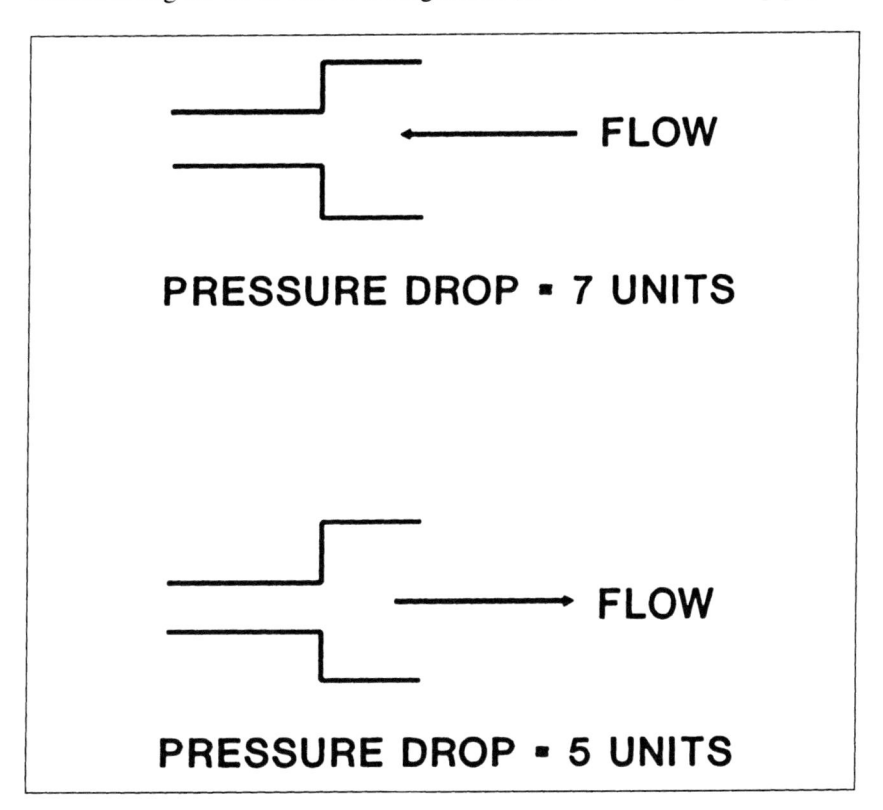

Figure 10-8. Pressure drop in enlargements and contractions.

gas flowing in one direction might have a pressure drop of 7 units. Flowing in the opposite direction, however, the pressure drop might be only 5

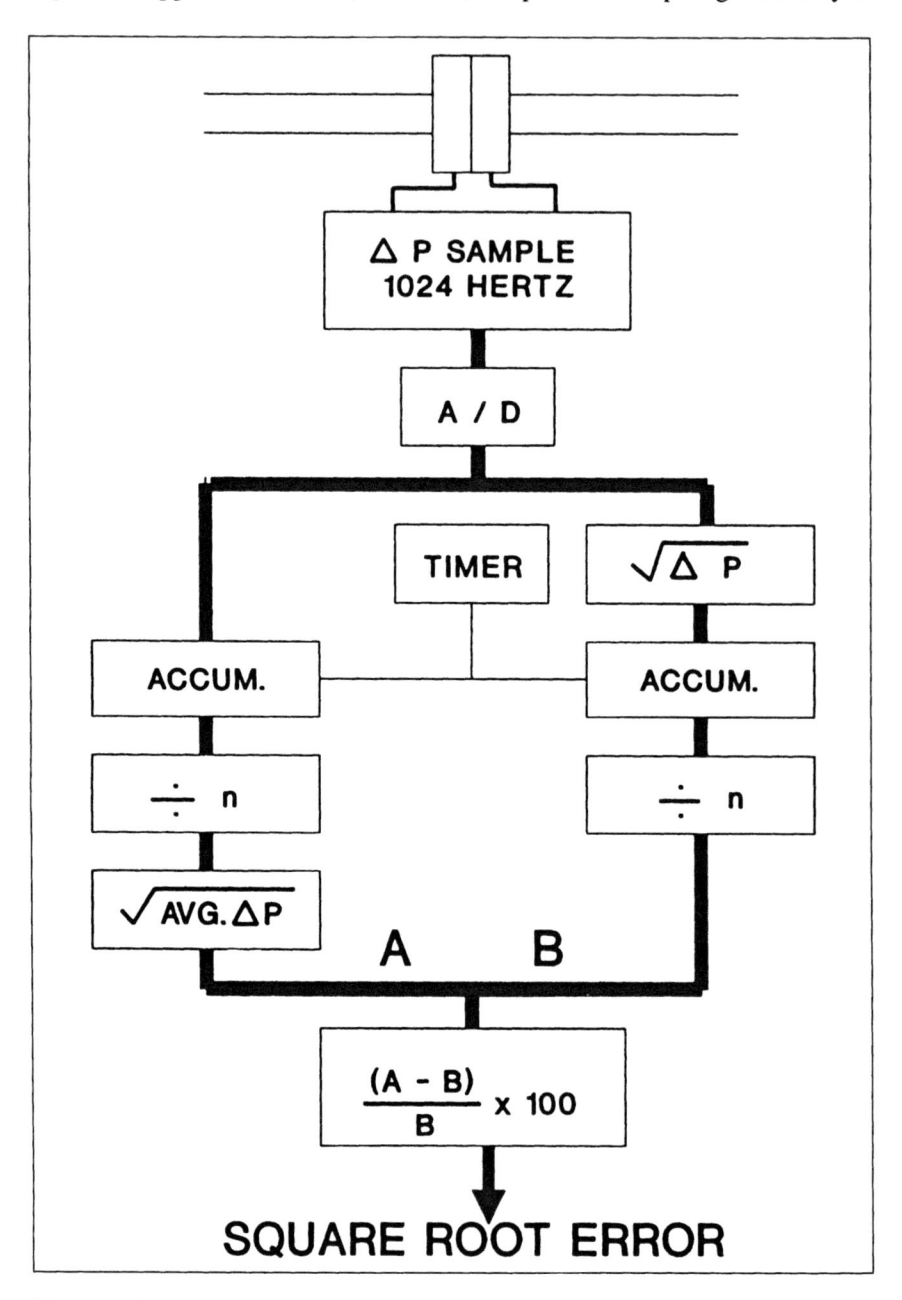

Figure 10-9. Information flow in the square root error indicator.

units. With pulsating flow in the gauge lines, this reversal of flow could cause a buildup of pressure in one gauge line and severely distort the true differential pressure.

Square Root Error Indicator. In 1984, the Pipeline and Compressor Research Council developed an instrument called the square root error indicator (SREI). Figure 10-9 is a schematic diagram showing the flow of information in this instrument. When connected to the taps of an orifice meter, the SREI can determine the amount of flow error due to pulsation that exists *at the orifice taps.* It does not measure error due to gauge line effects.

One of the questions frequently asked about the metering of pulsating flow concerns the use of electronic flow metering and whether it can correct for the effect of pulsations. At the present time, the fastest rate that the best commercially available field grade electronic transducers can sample is 4 times per second. By referring to Figure 10-7 it can be seen that this sampling rate is totally inadequate to define the wave shape of highly varying flow. At some time in the future the transducer technology may improve to the point where this will not be a limitation.

In storage facilities where a compressor is located near an orifice meter, the following steps are recommended:

1. Install a properly designed pulsation control system.
2. Keep the gauge lines between the orifice taps and the meter chart as short in length and as large in diameter as practical.

References

1. American Gas Association. *Gas Committee Report No. 3,* 1985.
2. Upp, E. L. *Fluid Flow Measurement.* Houston: Gulf Publishing Company, 1993.
3. Flanigan, Orin. *The Metering of Pulsating Flow: One Company's Case History.* Proceedings of the 27th Annual School of Gas Measurement Technology, 1992.
4. Sparks, C. R. *Pulsation and Transient-Induced Errors at Orifice Meter Installations.* Pipeline and Compressor Research Council Research Report 87-3, April, 1987.

CHAPTER 11

Dehydration

During the summer season gas is taken from the pipeline system and injected into storage. This gas that is injected is usually pipeline quality gas that has been dehydrated and has a low water content. The reservoir into which this gas is injected has some water in it. This water may be from an active water drive or it may be connate water that resides in a volumetric reservoir. In practically all cases, however, water is present in the reservoir. When the dry injected gas comes in contact with this water, part of the water vaporizes into the gas. The gas then becomes saturated with the water in the reservoir. Since the reservoir temperature may be significantly higher than the ambient temperature, the gas can hold significant amounts of water. During the withdrawal season this warm reservoir gas that is saturated with water will be withdrawn and exposed to quite cold ambient and ground temperatures. This change in temperature will cause part of the water in the gas to condense to a liquid.

Hydrates

When natural gas at high pressure is exposed to liquid water, the gas can physically combine with the water to form a hydrate. This ice-like material can form at temperatures significantly above 32°F. At very high pressures, hydrates have been observed in the neighborhood of 60°F. Because these hydrates only exist at high pressures, they are difficult to study, and because of this considerable information has yet to be learned about them. One observation is that these hydrates can form in a pipeline and block off the flow of gas. When this occurs the usual practice is to blow down the pipeline and reduce the pressure. This reduction in pressure causes the hydrates to decompose, allowing the flow of gas to resume. The preventative measure for this situation is to reduce the amount of water in the gas to the point where liquid water cannot form. This reduction in water content of the gas is achieved by dehydration. Dehydration is the partial or complete removal of water from a substance. For this discussion, dehydration will be limited to the partial removal of water from natural gas.

Water Content of Natural Gas

Considerable research has been done on the water content of natural gas at saturation. There are various charts that have resulted from this research. Typically these charts will consist of a graph with the gas temperature on the x-axis and water content of the gas at saturation on the y-axis. The water content is expressed as pounds of water per million standard cubic feet of gas. The parameters are lines of constant gas pressure. At low pressures the gas can hold much more water than at high pressures. At high temperatures the gas can hold more water than at low temperatures. One of the decisions the engineer must make is to determine what the water content of the dehydrated gas should be. In order to do this, it must be determined what the pressure of the gas will be and what will be the lowest temperature that the gas will reach. Once this is done, the water content charts may be used to determine the water content of the gas that must be achieved. As an example, assume that the gas going from the storage facility to the pipeline system is at 800 psig. The coldest temperature that the gas is expected to encounter will be 30°F. By using one of the water content charts, it can be seen that the water content of the gas at 800 psig and 30°F will be about 7.3 pounds per million cubic feet of gas.

For many years there has been an informal standard that pipeline quality gas must have no more than seven pounds per million cubic feet. This value is written into many gas purchase and sales contracts. In more recent years there has been a more conservative trend to require that the gas have a water content of no more than four pounds per million. This is certainly a safer standard and gives more latitude for error. As an example, if a dehydrator on the system begins to malfunction and put out non-specification gas, the remaining system gas may be able to mix with the poor quality gas and prevent hydrate troubles if the remaining dehydrators are producing four pounds per million gas. The decision as to whether to use a specification containing four pounds or seven pounds is a judgement determination. The four pound specification costs more to achieve, but it gives better protection.

Types of Dehydrators

In the distant past, gas dehydration was performed by the adsorption process. The dehydration vessels consisted of two or more pressure towers filled with solid pellets. These pellets were activated alumina or other material that had a very large internal surface area for their weight. The pellets had an affinity for the water vapor molecules, and the water vapor was adsorbed onto the pellets. After a period of time, the tower was taken out of dehydration service and the pellets were subjected to high temperatures. This high temperature and low pressure drove the water molecules out of the pellets. After cooling, the tower was ready to go back to dehydration service.

Solid bed adsorption type dehydrators are not used for natural gas dehydration to any extent any more except for special service or for units that have been in service for many years. The reason is installation and operation cost. Practically all new dehydrators put in natural gas service are glycol dehydrators. Glycol dehydrators have been around for many years. During their early years they were economical, but they could not perform dehydration as well as the solid bed units. They had limited dew point depression capability. In the past 30 years substantial advances have been made in glycol dehydration. Today a glycol dehydrator can be designed to achieve almost any dew point depression with practically any operating condition that may be encountered in the natural gas industry.

During this time the type of glycol used for dehydration has changed also. Originally ethylene glycol was used. Following this, diethylene glycol was used. Today triethylene glycol is used in most of the dehydrator installations in the gas industry.

Glycol Dehydrators

Cycle of Operation. Figure 11-1 shows a highly simplified diagram of a glycol dehydrator. The absorber consists of a pressure vessel that contains some means of placing the wet gas and the glycol in intimate contact. This is often done with bubble cap trays, although the column may be filled with ceramic packing when the diameter of the tower is small. A recent development is the use of a flooded tray, which combines the bubble cap tray and the ceramic packing. The gas enters the bottom of the vessel and flows upward. Glycol that has been stripped of essentially all its water content is introduced into the top of the tower and flows down through the tower by gravity. This counter-current process puts the purest glycol in contact with the driest gas. The dry gas leaves the top of the tower. The wet glycol that has absorbed water from the gas leaves the bottom of the tower.

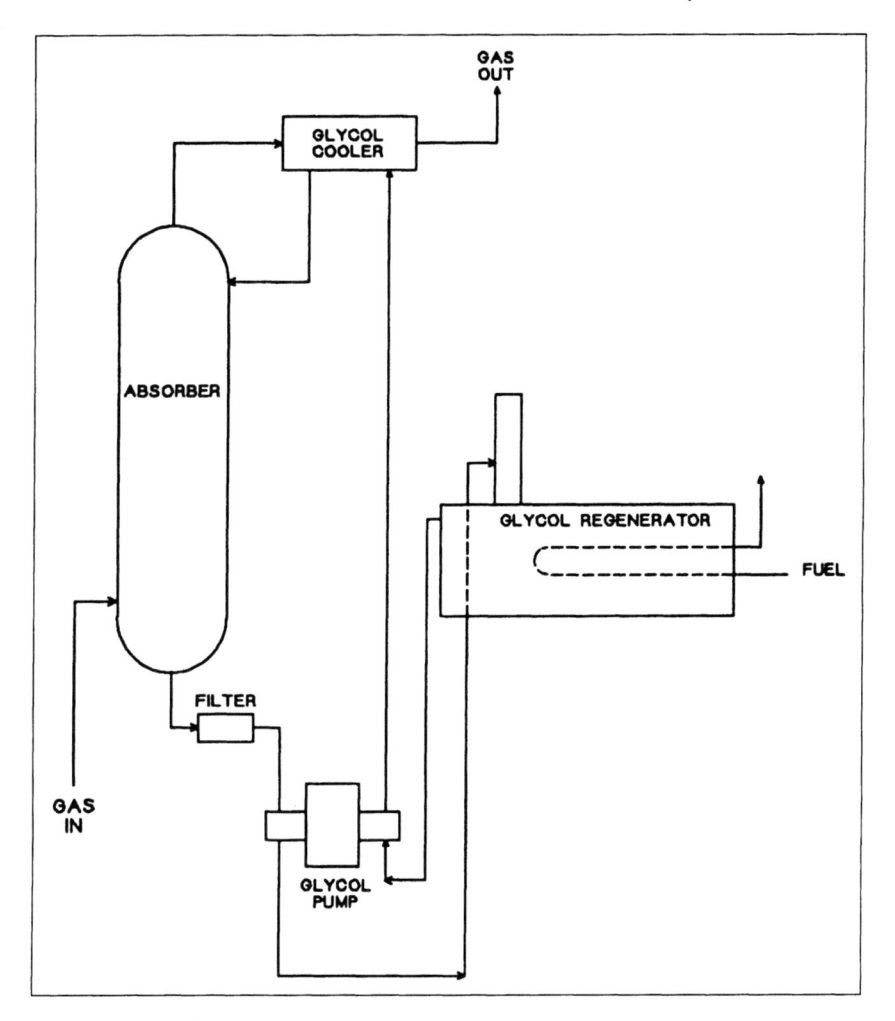

Figure 11-1. Schematic diagram of a glycol dehydrator.

The glycol leaving the bottom of the absorber contains water and is at high pressure. Often this high-pressure glycol is used as a source of power. It is sent to the power side of a piston type pump, where it expends its pressure by producing work. This same pump is used to pump up the lean, regenerated glycol from atmospheric pressure to the pressure of the absorber. The material going from the bottom of the absorber to the pump actually consists of glycol plus a small quantity of high-pressure gas. This natural gas assists in the pumping of the lean glycol and keeps the pump from being a perpetual motion machine. This pump also acts as a level control for the absorber. When the bottom of the absorber has too much glycol, only glycol goes to the power side of the pump. The pump operates much slower on glycol than it does on natural gas. Therefore the pump slows down, decreasing the amount of lean glycol going to the tower. Conversely, when the bottom of the tower runs out of glycol, mostly natural gas goes to the pump. The pump speeds up and sends more lean glycol to the tower.

The glycol-glycol pump is an important part of the dehydration system. In the past there have been problems with these pumps. They had high maintenance costs and reliability problems. These problems have been solved by putting a high quality glycol filter ahead of the pump. In some dehydrator installations an electrically driven pump is used instead of a glycol powered model. This is particularly true in larger installations where electric power is available and a stand-by generator is installed. In small field installations, however, the glycol powered pump is widely used.

From the pump the wet glycol goes to the glycol regenerator. This is an atmospheric pressure vessel where the glycol is heated by natural gas. The gas burner tube usually consists of a gas burner firing into a U-tube that goes into the regenerator. The flue gases then return through the U-tube and go up a flue stack. Heat is transferred from the flue gases to the glycol being regenerated through the U-tube. The wet glycol from the pump passes through the hot glycol in the regenerator to gain preheat. This preheated wet glycol is then introduced into the still column that is on top of the regenerator. This still column is packed with ceramic or other packing. The water that has been vaporized by the glycol being heated passes up the still column and goes to the atmosphere. The remaining wet glycol falls down into the regenerator. Here the glycol is further heated to the thermostat set point. The recommended temperature is usually 400°F. It is important to maintain this temperature. At lower temperatures the glycol is not sufficiently regenerated and the outlet gas will not meet the water content specifications. At higher temperatures the glycol will begin to decompose. This decomposition rate is quite low for temperatures only slightly above 400°F. There is always the possibility of a thermostat wandering off temperature or malfunctioning. It is prudent to keep a test thermometer near the dehydrator and check the regenerator temperature periodically.

When the glycol has been heated to 400°F in the reboiler, the glycol purity is about 99 percent. Glycol at this level of purity is adequate for some dehydration jobs. In many cases, however, a lower gas outlet water content is needed than can be obtained with this level of purity. This has been addressed in two ways. One method is by sparging. In this method, dehydrated natural gas is introduced into the bottom of the regenerator. This dry gas bubbles up through the glycol in the regenerator, allowing the small amount of water in the glycol to vaporize into the dry gas. This can produce a higher level of purity. In later years, a more efficient method of concentrating the glycol has been used. This is by the use of stripping gas. Instead of introducing the dry gas into the bulk of the regenerator, the stripping gas is contacted with the lean glycol leaving the regenerator. Because of this, the stripping gas is contacting only the purest glycol, not the glycol of average purity. With stripping gas the purity of the regenerated glycol can be raised to about 99.9 percent. The exact purity depends on the amount of stripping gas used.

The hot, lean glycol leaves the regenerator and goes to the pump. Here the glycol is pumped up to the absorber pressure. The hot glycol must be cooled down before entering the absorber. Various means are used to accomplish this. Regardless of what other means are employed, most all dehydration systems have a gas-to-glycol heat exchanger. Here the gas exiting the absorber is used to cool the glycol. The temperature of the glycol going to the absorber should not exceed 120°F. This is not a rigid value, but the hotter the glycol, the more glycol that will be lost due to the vapor pressure of the glycol at the higher temperatures. This glycol loss can be a significant factor at high glycol temperatures. Similarly, the lean glycol going to the absorber should not be too cold. The lower temperature limit is about 50°F. Below this temperature the glycol begins to gel, and the contact between the glycol and the gas in the tower is poor.

Design Economics. When soliciting bids on a dehydration system, the engineer usually furnishes the information on the anticipated operating conditions. The engineer also furnishes the specifications for the outlet gas. Although the vendor will design the dehydration system, the engineer should be aware of some of the trade-offs that are available in order to adequately evaluate the bids. As an example, there is a trade-off between glycol circulation rate and the number of trays (either actual or theoretical) in the absorber. Glycol circulation rate is often expressed in terms of gallons of glycol per pound of water removed. Table 11-1 shows one example of this trade-off. Two cases are shown. In both cases the gas conditions are the same, and the specified outlet water content is the same. The glycol purity is the same in both cases. Case 1 involves a tower that has six bubble cap trays. Case 2 is for a tower that has ten bubble cap trays. In order to achieve the same outlet water content for the gas in both cases, the glycol circula-

Table 11-1
Effect of Number of Trays in Tower
on the Glycol Circulation Rate

	Case 1	Case 2
Gas Volume – MMcfd	10	10
Gas Pressure – psig	600	600
Gas Temperature – °F	120	120
Gas Inlet Water Content – #/MMcf	155	155
Gas Outlet Water Content – #/MMcf	7	7
Glycol Purity – Percent	99.8	99.8
Water Removed – #/MMcf	148	148
Total Water Removed – #	1,480	1,480
Number of Trays	6	10
Stripping Gas – cf/gal. of glycol	2.0	2.0
Glycol Circulation Rate Required – gal./# water	4.00	1.75
Glycol Circulation Rate – gallons per day	5,920	2,590
Regenerator Rating Required – Btu per hour	1,750,000	500,000

Source: Clawson [1]

tion rate for the tower with six trays must be 4.00 gallons of glycol per pound of water removed while in the case of the tower with ten trays, the circulation rate must be only 1.75 gallons per pound of water removed. Because Case 1 requires a higher circulation rate, it also requires a larger regenerator and a higher annual fuel consumption.

In this comparison, Case 2 will probably require the largest investment. This is because the high-pressure tower is larger. The key here is to determine which of the cases will have the lowest annual cost, including fixed costs on the investment. If low-cost fuel gas is available, Case 1 may be the best choice. If fuel gas is moderate or high priced, Case 2 may be the best choice. It is important, however, that first cost not be the only deciding factor.

A similar but less dramatic trade-off is available between absorber tower diameter and tray spacing. A tower that has a tray spacing of 26 inches or 30 inches instead of 22 inches will have a higher allowable velocity than a tower that has a tray spacing of 22 inches. This extra tray spacing helps prevent the carryover of glycol from one tray to the next and thus allows a higher gas velocity. In certain situations this increased velocity can cause a decrease in tower diameter and may save investment cost.

Volume Turndown. One of the concerns when dehydrators are used in storage service is the range of volumes. A storage facility may be turned on to withdraw 50 MMcfd in the early or late part of the winter season, and then be turned on to deliver 250 MMcfd during the colder part of the winter. This fairly wide range of flow rates can sometimes pose a problem for a single dehydrator. An absorber with bubble cap trays may have a maximum turndown ratio of 5. The turndown ratio is defined as the ratio of the maximum flow rate to the minimum flow rate during which the tower will perform adequately. An absorber that is packed with ceramic or other material may have a turndown ratio of 3. At lower ratios than this the gas has a tendency to channel up through the packing and good contact is not achieved between the gas and the glycol. Turndown ratios in the neighborhood of 100 have been claimed for the flooded tray type of absorber. Regardless of what type of tower is used, it is often desirable to use multiple dehydrator units in a large storage facility.

These multiple storage units can solve the problem of widely varying flow rates. They do introduce, however, the minor problem of equalization of flow between absorbers. It is usually desirable to have some type of flow measurement device on the inlet or outlet of each absorber to monitor the flow rate to the absorber. This device could be a pitot tube or a non-specification orifice. A butterfly valve should be provided to manually regulate the flow rate to each absorber.

Potential Problems. The absorber towers are usually plumb (vertically straight) when they are installed. It is good practice to check the tower or towers for plumbness periodically. If a tower leans, the flow of glycol across the trays will not be even and good gas-glycol contact will not be achieved.

Glycol purity is one of the keys to adequate dehydration. When the outlet gas water content is higher than specification, one of the first places to look is the temperature thermostat on the glycol regenerator. For triethylene glycol this temperature should be kept at 400°F. A good quality test thermometer should be available to check the glycol temperature. The still column can also be a source of trouble under certain conditions. If the gas flow rate to the absorber drops considerably below the design value, there may not be enough water being boiled out of the wet glycol to keep the still column warm. When this occurs, the water vapor rising up the still column condenses and runs back down the column instead of going out the column as a vapor. This condensed water goes back into the glycol being regenerated and dilutes it. This prevents the regenerator from achieving the 99 plus percent purity required, and the gas exiting the absorber contains more water than it should. If the lower gas volume is a temporary condition, the still column can be insulated so that the small amount of heat being generated is retained. If the lower gas volume is permanent, it may be desirable to put on a smaller diameter still column. There is also the possibility of injecting too much stripping gas into the regenerator. When this occurs the volume of stripping gas overloads the still column and allows some of the glycol to be carried out as droplets.

When a dehydration system is first started up in a new storage facility, the glycol should be checked about once a week for contaminants in the glycol, particularly chlorides. Salt water is often present in the gas withdrawn from storage. If the separation equipment is not entirely adequate the salt can find its way into the glycol. If the accumulation of salt in the glycol becomes great enough, it can coat the fire tube in the regenerator. This can cause overheating and eventual failure of the fire tube. After the weekly check for contaminants in the glycol has proved negative for several weeks, this check can be extended to three-month and then to six-month intervals.

Dew Point Measurement

The standard method for determining the quantity of water vapor in gas is to determine the dew point of the gas. Once the dew point is determined, there are a variety of graphs and charts that will convert the dew point into the water content in pounds of water vapor per million standard cubic feet of gas. Several devices measure the dew point of natural gas. The device that has come to be accepted as the standard is the U.S. Bureau of Mines dew point tester. This device consists of a pressure chamber capable of holding natural gas at high pressure. One side of the chamber consists of a flat mirror, and the opposite side consists of a transparent optical port. The gas at its pipeline pressure is introduced into the chamber and is allowed to flow at a low rate through the chamber and is vented to the atmosphere. The mirror on one wall of the pressure chamber is cooled by a refrigerant that is exposed to the back of the mirror. This refrigerant is usually propane that is introduced as a liquid into the cooling chamber. The vaporization of the propane liquid provides the cooling necessary. A mercury thermometer measures the temperature of the mirror that is being cooled. The temperature of the mirror is gradually lowered by introducing the propane refrigerant. At some point, the temperature of the mirror will get low enough such that the water vapor in the gas will condense on the mirror. This is observed through the transparent optical port. The temperature at which the first condensation of moisture is detected on the mirror is the dew point of the gas.

Some degree of skill and practice is necessary to successfully operate this dew point tester. The main skills of successful operation are the ability to control the rate of temperature decrease of the mirror and the ability to detect the moisture when it forms on the mirror. In running dew points on field gas from a producing well, there can be significant amounts of heavy hydrocarbons in the gas. These heavy hydrocarbons can condense on the mirror before the water dew point is reached. It is important to be able to detect the difference between a water dew point and a hydrocarbon dew point. Gas withdrawn from storage usually does not have significant amounts of heavy hydrocarbons in it. There have been cases, however, where pipeline quality gas was injected into an oil or gas-condensate reservoir that still had some liquid hydrocarbons residing in the gas sand. These heavy hydrocarbons would vaporize into the gas phase and could pose a problem during the withdrawal cycle.

It is also important to be aware of the atmospheric conditions surrounding a dew point tester when it is in use. Consider the case where a source of gas contains gas that has an actual dew point of 40°F and a gas temperature of 40°F. The atmospheric temperature is 25°F, and the gas source is connected to the dew point tester with a length of one-quarter inch copper tubing. The small diameter tubing and the low flow rate of the gas to the tester causes the gas to be cooled to the atmospheric temperature of 25°F. This causes water to condense out of the gas, so that the gas reaching the tester has a dew point of 25°F. The dew point tester will then read a dew point of 25°F, while the dew point of the gas at the source is 40°F. In a case like this, some means must be used to prevent the cooling down of the gas before reaching the tester.

There are various other devices that are used to measure dew points. One disadvantage of the Bureau of Mines unit is that it is a batch type unit and it must be manually operated. There is a real need for a fully automatic, continuous device that will reliably measure dew points with reasonable maintenance. Several devices have been proposed for this type of service. These include chemical, electronic, and combinations of the two. These have been used with varying success over the years.

Underground Gas Storage Facilities

Water Content Charts

Once the dew point of the gas has been obtained, this information must be converted into water content by the use of a dew point graph or table. The typical dew point graph consists of a graph on which the x-axis is temperature, the y-axis is water content of the gas in pounds of water per million standard cubic feet of gas, and lines of constant pressure are the parameter. There are a number of these tables and graphs. These include the following:

McKetta and Wehe [2]
Institute of Gas Technology [3]
Deaton and Frost [4]
McCarthy, Boyd, and Reid [5]
Handbook of Natural Gas Engineering [6]
Gas Conditioning and Processing [7]

Each of these give good answers, but each set of values is somewhat different from any other set. In an attempt to compare the values from each source, a series of pressures and temperatures were selected. The pressures selected were 14.7, 100, 300, 500 and 800 psia. The temperatures selected were −40, −20, 0, 20, 40, 80, and 120°F. This gave a matrix of 35 points for comparison. The water content values for each of these points was carefully evaluated for each source.

In order to compare the values from the various charts some standard is needed. Since no standard was available, the point values from all the charts were averaged and that average set of values was used as a standard. For each combination of pressure and temperature the water content values from all the sources were averaged. The water content value from each source was then compared to this average, and the percent deviation for each source was obtained for that point. This deviation was converted to an absolute value to prevent a large negative deviation and a large positive deviation from canceling each other. This procedure was followed for each of the temperature-pressure points. For each source, the absolute deviations for each point for that source were averaged and compared. The result of this comparison was:

Source	Average of Absolute Deviations
McKetta and Wehe	3.4%
Institute of Gas Technology	5.3%
Deaton and Frost	9.2%
McCarthy, Boyd, and Reid	2.7%
Handbook of Natural Gas Engineering	4.4%
Gas Conditioning and Processing	3.9%

The tabulation shows that there is some degree of difference between the various sources of values. There is a rather large difference, however, for the values of Deaton and Frost. Because it was felt that this data may skew the averages, the values of Deaton and Frost were eliminated. This does not imply that the values of Deaton and Frost are wrong. It does mean that they are different from the other sources. The averages were then recalculated without the Deaton and Frost values and the same comparison made. The result was:

Source	Average of Absolute Deviations
McKetta and Wehe	3.3%
Institute of Gas Technology	2.4%
McCarthy, Boyd, and Reid	2.2%
Handbook of Natural Gas Engineering	4.1%
Gas Conditioning and Processing	3.2%

This comparison shows that all of the sources are relatively close to the average. The values of McCarthy, Boyd, and Reid have the closest agreement with the averages by a small margin. All this does not mean that the values of McCarthy, Boyd, and Reid are most correct. There is no basis for making that determination. It does mean, however, that the values of McCarthy, Boyd, and Reid have better agreement with all the other charts than any other chart has. This can be an advantage.

Tabulating Water Content Data

The reading of water content values from a dew point chart is at best a chore. Further, it is not an error-free process. A much more desirable alternative would be a set of tabular values. Due to the nature and nonlinearity of the dew point curves, the curve fitting of data points presents a challenge. Behr [8] has suggested the following form of an equation for correlating the values:

$$\ln W = A + B \times \left(\frac{1}{T}\right)^2 + C \times \left(\frac{1}{T}\right)^3 + D \times (\ln P) + \\ + E \times (\ln P)^2 + F \times (\ln P)^3 + G \times \left(\ln \frac{P}{T}\right)^2 + H \times \left(\ln \frac{P}{T}\right)^3 \qquad (11\text{-}1)$$

W = water content - #/MMScf
P = pressure - psia
T = dew point temperature - °Rankine
A, B, C, D, E, F, G, H = constants to be evaluated

Using the water content values of McCarthy, Boyd, and Reid the constants were evaluated using a regression analysis technique. The values obtained were:

A = 13.293150
B = −4,261,306
C = 662,351,880
D = 1.8362996
E = −0.36872304
F = 0.016793440
G = 0.1751796
H = −0.0086638038

The tabulated values can be put in a format similar to that shown in Table 11-2. The pages are 5.5 inches wide and 8.5 inches high. When put in a loose-leaf binder, the binder will conveniently fit into a truck glove compartment or a brief case. There is no problem with opening this volume up in the wind or rain. Further, the values read are not subject to astigmatism or myopia.

Table 11-2
Water Vapor Content of Natural Gas in Pounds per MMcf

Pres. psig	\multicolumn{10}{c}{Temperature – Degrees Fahrenheit}
	90	91	92	93	94	95	96	97	98	99
250	139	143	147	151	156	161	165	170	175	180
255	136	140	145	149	153	158	163	167	172	177
260	134	138	142	147	151	155	160	165	170	175
265	132	136	140	144	149	153	158	162	167	172
270	130	134	138	142	146	151	155	160	164	169
275	128	132	136	140	144	148	153	157	162	167
280	126	130	134	138	142	146	151	155	160	164
285	124	128	132	136	140	144	148	153	157	162
290	123	126	130	134	138	142	146	151	155	160
295	121	125	128	132	136	140	144	149	153	157
300	119	123	127	130	134	138	142	146	151	155
305	118	121	125	129	132	136	140	144	149	153
310	116	120	123	127	131	135	138	143	147	151
315	115	118	122	125	129	133	137	141	145	149
320	113	117	120	124	127	131	135	139	143	147
325	112	115	119	122	126	129	133	137	141	145
330	110	114	117	121	124	128	132	135	139	143
335	109	112	116	119	123	126	130	134	138	142
340	108	111	114	118	121	125	128	132	136	140
345	106	110	113	116	120	123	127	131	134	138
350	105	108	112	115	118	122	125	129	133	137
355	104	107	110	114	117	120	124	128	131	135
360	103	106	109	112	116	119	122	126	130	134
365	102	105	108	111	114	118	121	125	128	132
370	100	103	107	110	113	116	120	123	127	131
375	99	102	105	109	112	115	118	122	125	129
380	98	101	104	107	111	114	117	121	124	128
385	97	100	103	106	109	113	116	119	123	126
390	96	99	102	105	108	111	115	118	122	125
395	95	98	101	104	107	110	114	117	120	124
400	94	97	100	103	106	109	112	116	119	122
405	93	96	99	102	105	108	111	114	118	121
410	92	95	98	101	104	107	110	113	117	120
415	91	94	97	100	103	106	109	112	116	119
420	91	93	96	99	102	105	108	111	114	118
425	90	92	95	98	101	104	107	110	113	117
430	89	92	94	97	100	103	106	109	112	115
435	88	91	93	96	99	102	105	108	111	114
440	87	90	93	95	98	101	104	107	110	113
445	86	89	92	94	97	100	103	106	109	112
450	86	88	91	94	96	99	102	105	108	111
455	85	87	90	93	96	98	101	104	107	110
460	84	87	89	92	95	97	100	103	106	109
465	83	86	89	91	94	97	99	102	105	108
470	83	85	88	90	93	96	99	101	104	107
475	82	85	87	90	92	95	98	101	104	107
480	81	84	86	89	92	94	97	100	103	106
485	81	83	86	88	91	93	96	99	102	105
490	80	82	85	87	90	93	95	98	101	104
495	79	82	84	87	89	92	95	97	100	103

References

1. Clawson, J. S. Private Communication. 1994.
2. McKetta, John J. and Wehe, Albert H. "Use This Chart for Water Content of Natural Gases." *Petroleum Refiner,* Aug., 1958, pp. 153–154.
3. Bukacek, R. F. "Equilibrium Moisture Content of Natural Gases." *Institute of Gas Technology Research Bulletin 8,* Nov., 1955.
4. Deaton, W. M. and Frost, E. M. Jr. "Gas Hydrates and Their Relation to the Operation of Natural-Gas Pipe Lines." *U.S. Bureau of Mines Monograph 8.*
5. McCarthy, E. L., Boyd, W. L., and Reid, L. S. Trans. AIME, 189, 1950, pp. 241–242.
6. Katz, D. L., *Handbook of Natural Gas Engineering.* New York: McGraw Hill, 1959.
7. Gas Conditioning and Processing, Volume 2.
8. Behr, W. R. "Correlation Eases Absorber-Equilibrium-Line Calculations for TEG-Natural Gas Dehydration." *Oil and Gas Journal,* Nov. 7, 1983, pp. 96–98.
9. Martin, L. K. and Flanigan, Orin. "A Comparison of Water Vapor Content Charts for Natural Gas." Unpublished internal report. January, 1984.

CHAPTER 12

Compressors

When compressors are used in underground storage service, the application is somewhat unique and fairly demanding. For a portion of the year the compressors are handling a relatively small amount of gas and are compressing it at a relatively high compression ratio. For another portion of the year, the service is just the opposite. The units are handling large volumes of gas at somewhat lower compression ratios. This dual application poses some interesting challenges in the selection of equipment and in the operation and maintenance of that equipment.

Injection versus Withdrawal Requirements

The volume of gas that must be handled during the withdrawal cycle usually determines the number and size of compressor cylinders. When these same compressor cylinders are put into service during the injection cycle, the compression ratio is usually considerably higher. This high compression ratio with the large amount of compressor cylinder displacement tends to overload the engine, requiring more horsepower than the engine rating. Clearance pockets on the compressor cylinders can help alleviate this problem. In many cases, however, it is not physically possible to put enough clearance pockets on the cylinders to reduce the horsepower. In those cases some of the compressor cylinders must be unloaded with the use of valve lifters or some similar device. The valve lifter renders the valve ineffective so that the gas, usually suction gas, enters the cylinder during the suction stroke and is discharged back to the suction manifold during the discharge stroke. Thus the cylinder develops no effective horsepower because no gas is being pumped from a low pressure to a higher pressure.

These cylinder unloaders solve the problem of overloading the engine, but they introduce a second problem. Although the compressor cylinder is generating no useful work, it is still generating some work. During the suction stroke there is some pressure drop across the suction valve and the inlet passage even though the suction valve is lifted. Similarly, during the discharge stroke there is some pressure drop across the suction valve and flow passage. The cylinder must perform work to overcome these pressure drops. This causes a rise in the temperature of the gas. Since gas is not flowing through the cylinder, there is no cooling effect due to new, cool gas entering the cylinder. This causes the cylinder temperature to rise. It also causes the temperature of the gas in the suction manifold to rise. Unless this heat is dissipated it can cause excessive temperatures in the unloaded cylinder and increased horsepower usage due to the suction gas temperature being higher than normal. Water cooled compressor cylinders should be considered for a situation such as this.

During the withdrawal cycle the compressor operates at relatively low compression ratios. During the injection cycle the compression ratios may be quite high. It is not unreasonable to have a maximum reservoir pressure of 2,500 psia. If pipeline gas is available at 800 psia, this results in a compression ratio of 3.125. Although this magnitude of compression ratio has been successfully handled with a single compression stage, many operators feel that the compression ratio should be limited to 2.50, and that two stages of compression should be used for ratios above this. Other than excessive rod load, the usual limiting factor is the temperature of the discharge gas. A reasonable maximum temperature for this discharge gas is about 200°F. There is often some liquid in the form of mist or droplets that enters the compressor cylinder even though the gas has been through an inlet scrubber. At temperatures above about 200°F the vapor pressure of these liquids increases to the point that some flashing takes place in the cylinder, leaving a deposit of whatever material was dissolved in the liquid. This can cause a buildup of solids on the cylinder wall, which will cause excessive wear.

Lubrication Options

There are three lubrication options in which compressor cylinders may be operated. The historic method was to operate with full lubrication. Very few companies operate their newer compressors fully lubricated at the present time, although some of the older vintage compressors are operated in this manner. The other options are non-lube and mini-lube. Under the non-lube option, no lubrication is added to the compressor cylinder. Teflon and other similar materials are used where appropriate on moving surfaces. When the non-lube option is used, some operators recommend rotating the compressor pistons 90 degrees every year. With the mini-lube option, the same friction-resistant materials are used, but a minimum amount of lubrication is added. The relative advantages of these two options is not completely clear. There has been good experience with both. There does appear to be general agreement, however, that operation between the mini-lube option and the fully lubricated option can cause maintenance problems.

Two-cycle versus Four-cycle Engines

There are varying views concerning two-cycle versus four-cycle engines. This variation in views is likely to continue. The four-cycle engines have historically had a wider load range than the two-cycle units. This was because the reliability of firing in the two-cycle engines was not good at lower operating loads. With the advent of lean burn engines and the stratified charge method of firing, however, the stratified charge essentially acts as a torch. This has improved the lower torque range of the two-cycle units considerably.

Control Technology

As the compressor-engine technology has developed, some control methods and philosophies have also developed. One of these is the fuel-air ratio controller. This instrumentation is now standard on all integral units of any size. Another is the torque controller. Many years ago a compressor load curve was used to determine the correct clearance pocket settings for the particular operating condition. The operator would take the suction and discharge pressures, refer to a load curve to determine the clearance pocket settings, and operate the appropriate pockets. This procedure was necessary to keep the engine fully loaded, but not in excess of its rated horsepower. Now this procedure is done automatically. Sometimes a load calculation is made by computer, and the pockets are operated hydraulically from this information. When the engine is in good condition, this method works well. When the engine is "sick," however, this method can put more load on the engine than the engine can take. Another method is to use the measured fuel rate to the engine as a measure of torque. This method has the advantage of effectively taking into account the condition of the engine.

Inlet Separator

Every compressor installation should include a good separation system ahead of the compressors to take out any solid and liquid material. A filter separator is usually used for this purpose. The filter separator consists of a pressure vessel that has some type of separation volume. It also has one or more metal forms in the shape of long narrow cylinders with holes in each one. A fabric "sock" is placed over the form and the inlet gas is allowed to pass through the sock. Conventional type units were set up so that the gas flowed from the outside of the sock to the inside, and the metal form provided a path for the gas to exit the filter. This conventional type unit had the advantage that it offered the maximum amount of surface for the collection of solid particles. A more recent type of filter separator is called a reverse flow unit. This reverse flow causes the gas to enter through the core of the sock and to flow outward through the filter material. This type of unit has the characteristic that the minimum gas velocity occurs as the gas is leaving the filter material. This has an advantage in that any liquid that is caught in the filter material is more likely to coalesce and drop to the bottom of the pressure vessel due to the lower velocity, whereas a higher velocity might cause it to re-enter the gas stream. In either type of unit the filter socks must be changed periodically when they build up pressure drop.

Fin Fan Coolers

In a reciprocating compressor facility there is a need for cooling. This cooling requirement includes the engine jacket water, the engine oil, the outlet gas, and sometimes the compressor cylinder jacket water. The common method of cooling is through the use of air-to-liquid coolers, usually referred to as fin fan coolers. These fin fan coolers may be forced draft or induced draft units, with the fan moving the air located either below or above the finned tubes. The choice of induced versus forced draft is a matter of preference. Of greater importance is adequate sizing of the coolers. In the southern region of the country these units are often designed for an ambient temperature of 100°F. Although this ambient temperature design point is valid, the temperature of the air entering the coolers may be greater than 100°F. Part of this is due to radiation from the concrete and gravel in the vicinity of the coolers. Part of it may be due to an inevitable amount of recirculation of the hot exit air. For all of these reasons the actual air temperature entering the coolers may approach 103 to 105°F. These few degrees may compromise the performance of the coolers on hot days.

The fans for the coolers may be driven by an electric motor or by a hydraulic drive operating on power from the compressor engine. Both methods work satisfactorily. If electric motors are used, a standby electric generator must be available in the event of electric power failure. The hydraulic drive off the compressor engine offers the advantage that no additional environmental measures need be taken due to a standby electric generator. The disadvantage is that a small portion of the engine's total horsepower is not available for compression. There is a potential problem with the hydraulic drive system. The hydraulic fluid at moderate temperatures has a reasonable viscosity and it operates well in the pump and in the hydraulic motors. When the temperature of the hydraulic fluid increases, the viscosity of the fluid goes down. This results in slippage in the pump and in the motors. This can cause a decrease in the effectiveness of the system in very hot weather. A hydraulic fluid cooler should be installed in the fin fans to keep the temperature of this fluid such that the viscosity is acceptable.

Preventative Maintenance

Integral engine-compressors are major pieces of equipment. Reliability is a key factor in any component of an underground storage facility. Maintenance is necessary on any moving piece of machinery. A good preventative maintenance schedule is necessary for this reliability. Compressor analyzers will tell when something is wrong or broken in the compressor; they will not tell when something is about to break. There are several good analyzers on the market. One recent development in this field is a joint venture between the Gas Research Institute and the Pipeline and Compressor Research Council, which produced Compressor Diagnostic Software. It appears to have considerable promise as a diagnostic tool.

Ideally, the compressor cylinders should be examined monthly by an analyzer. The power cylinders should be examined annually with a bore scope. It is impractical to perform a complete teardown of the engine-compressor unit every year. There can be, however, an annual look at some part of the internals. For example, one main bearing might be examined each year. A total overhaul might be performed every 150,000 hours or so.

Compressor Models

The modeling of an underground storage facility on a computer requires that a compressor model be available. These compressor models may be as simple as a single equation using suction pressure, discharge pressure, and flow rate to calculate the horsepower required. They may be

152 *Underground Gas Storage Facilities*

as complex as a series of equations that model the individual characteristics of a specific brand and model of compressor and allow computation of flow and horsepower taking into account the individual cylinder clearance and the compressor speed. While the latter type of model is quite beneficial in some situations, there are many cases where a good, reliable single equation representation of a compressor is not only adequate but desirable.

In the not-too-distant past it was difficult to measure the horsepower of a reciprocating compressor in a field installation with reasonable certainty and repeatability. With the development of analyzers as they are today, this situation has changed. At present, analyzers can determine the gas horsepower of a compressor with good certainty and repeatability. Unfortunately the progress in the area of predicting horsepower requirements has lagged behind this effort. Compressor manufacturers have computer programs for the determination of horsepower developed by their equipment and the flow throughput resulting from this horsepower. These programs are somewhat complex and are not easily integrated into a system model for the entire storage facility. Needed are some rather simple equations that will predict horsepower within some reasonable uncertainty.

Appendix B shows the development of the theoretical equation for compressor horsepower. This equation is:

$$\frac{HP}{Q \times T_1 \times Z_1} = 0.08571 \times \left(\frac{k}{k-1}\right) \times \left[\left(\frac{P_2 \times Z_1}{P_1 \times Z_2}\right)^{\frac{k-1}{k}} - 1\right] \qquad (12\text{--}1)$$

HP = horsepower
Q = gas volume in MMScfd
P = pressure in psia
T = temperature in °Rankine
Z = compressibility of gas
k = ratio of specific heats of gas

This equation is theoretical and assumes 100 percent efficiency in all phases of the compression with no friction losses in either the gas flow or the mechanical parts. Experience has shown that the actual horsepower required is considerably in excess of this theoretical equation.

This theoretical equation is also nonlinear due to the exponential term. Angulo [1] has suggested a linear modification of the equation to represent a particular compressor over a limited range of compression ratios. The correlation equation is:

$$\frac{HP}{MMScfd} = \left[A \times \left(\frac{P_D}{P_S}\right) + B \right] \times T_S \times Z_S \qquad (12\text{-}2)$$

where A and B are correlation constants that must be evaluated from data. Angulo achieved good correlation with this equation for a specific compressor for compression ratios between 1.2 and 1.6. The correlation was obtained by regression analysis of analyzer data from the compressor.

Flanigan [2] has suggested another approach. Equation 12-1 may be rearranged to the form:

$$\frac{HP \times (k-1)}{Q \times T_S \times Z_S} = 0.08571 \times \left[\left(\frac{P_D \times Z_S}{P_S \times Z_D}\right)^{\frac{k-1}{k}} \right] - 0.8571 \qquad (12\text{-}3)$$

The term Z_S/Z_D approaches a value of 1.0 because of the temperature rise that occurs during compression. If this adjustment is made and the terms 0.0857 are replaced by correlation factors "a" and "b", the equation becomes:

$$\frac{HP \times (k-1)}{Q \times T_S \times Z_S \times k} = a + b \times \left(\frac{P_D}{P_S}\right)^{\frac{k-1}{k}} \qquad (12\text{-}4)$$

This equation has the general form of Y = a + bX. If the term:

$$\frac{HP \times (k-1)}{Q \times T_S \times Z_S \times k}$$

is plotted against the term:

$$\left[\frac{P_D}{P_S}\right]^{\frac{k-1}{k}}$$

the result should be a straight line with the intercept "a" and the slope "b". In order to test this, the manufacturers' curves for horsepower and throughput for specific compressors were used to obtain data for plotting. Data points were obtained for each compressor, and the data for each compressor plotted in a very tight correlation pattern. Figure 12-1 shows a typical plot. From this information the constants "a" and "b" were evaluated. The results are shown in Table 12-1.

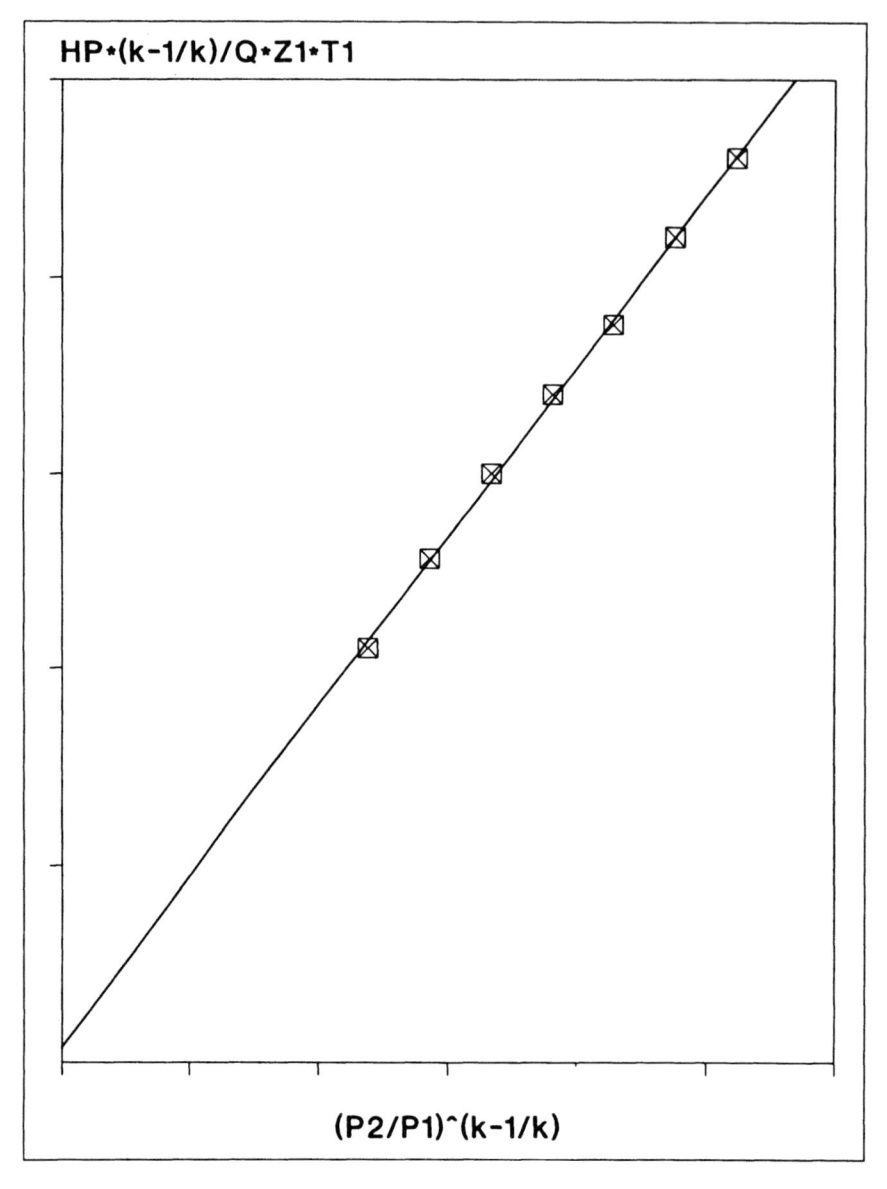

HP*(k-1/k)/Q*Z1*T1

(P2/P1)^(k-1/k)

Figure 12-1. Typical correlation of horsepwer data for compressors.

Table 12-1
Horsepower Correlation Factors for Various Types of Integral Compressors

Comp. Type Number	Factor a	Factor b	Percent Over Theoretical
1	(0.1116)	0.1119	30.6
2	(0.1112)	0.1112	29.8
3	(0.0986)	0.0986	15.1
4	(0.0985)	0.0996	16.2
5	(0.0997)	0.0997	16.3
6	(0.1044)	0.1052	22.8
7	(0.0896)	0.0904	5.5
8	(0.0911)	0.0915	6.8
Average	(0.1006)	0.1010	17.9
Theoretical	(0.0857)	0.0857	

A total of eight units were evaluated representing eight different models of integral compressors from three different manufacturers and of widely different dates of manufacture. The information contained in the results allowed several conclusions to be drawn.

1. The magnitude of the constants "a" and "b" is a measure of the efficiency of the compressor unit. The larger the value of these constants is, the more horsepower is required and the less efficient the compressors are.
2. The constants "a" and "b" are essentially equal (with opposite sign) to each other for each compressor. This follows the theoretical pattern.
3. Larger horsepower compressors were in general more efficient than smaller horsepower compressors.
4. Later model compressors were in general more efficient than earlier model compressors.
5. There can be significant differences in efficiencies of compressors of similar size and vintage.
6. All of the compressors require significantly more horsepower than indicated by the theoretical equation.

The method illustrated by Figure 12-1 is a means of simplifying the manufacturers' curves to the point that the information can be used in a computer model of the storage facility. If the type of compressor is unknown, a value of 0.1 can be used for the constants "a" and "b" for single stage integral compressors. It is suggested that constants of 0.105 be used for generic high-speed units and constants of 0.110 be used for two-stage units. If the specific model and brand of compressor is known, the manufacturers' data may be used to evaluate these constants. This method can also be used as an aid in evaluating compressors during the purchasing phase.

REFERENCES

1. Angulo, Manual G. *Empirical Horsepower Equations.* Proceedings of the 8th PCRC Reciprocating Machinery Conference, Sept. 20–23, 1993.
2. Flanigan, Orin. *Definition of a Horsepower Equation for Use in Flow Studies.* Unpublished internal report, 1982.
3. Pipeline and Compressor Research Council. *Field Measurement Guidelines—Compressor Cylinder Performance Summary,* Technical Report 84-10A, 3rd revision, Feb. 1990.
4. Osmon, Paul. Private communication.

CHAPTER 13

Estimating Deliverability of Producing Wells

Deliverability may be defined as the maximum rate at which gas may be produced at any point in time from a well, reservoir, field, or producing property. When any reservoir first begins producing, the deliverability is relatively high. As the reservoir is produced, the cumulative production causes a decrease in the total gas in place. This decrease in the total gas in place causes the reservoir pressure to decrease. This in turn causes a decrease in deliverability. In a producing field it is important to be able to predict what the deliverability of the field will be at some future time when a given amount of gas has been produced. This is necessary in order to know what quantity of gas will be available on a given day from field supply. Once the field supply deliverability is known and the deliverability from other sources is known, these quantities are matched against the load curve of customer requirements. The difference is the amount of gas that must be supplied on a daily basis from storage. This type of analysis is necessary in order to make 1-year and 5-year forecasts of supply and requirements.

Historical Deliverability Forecasting

Pressure Decline Curve. Estimating field deliverability has historically been the province of the petroleum engineer. The petroleum engineer used a three-part procedure to make these estimates. The first part of the procedure was to obtain a pressure decline curve of each reservoir on the system. Figure 13-1 shows a typical pressure decline curve for a reservoir. This curve shows how the reservoir pressure varies as the gas is produced from the reservoir. Obtaining this pressure decline curve is not always a simple task. For new fields there is very little history to plot. This makes necessary a large extrapolation of the existing data into the future. This forecasting can involve considerable error. Even with the older producing

158 *Underground Gas Storage Facilities*

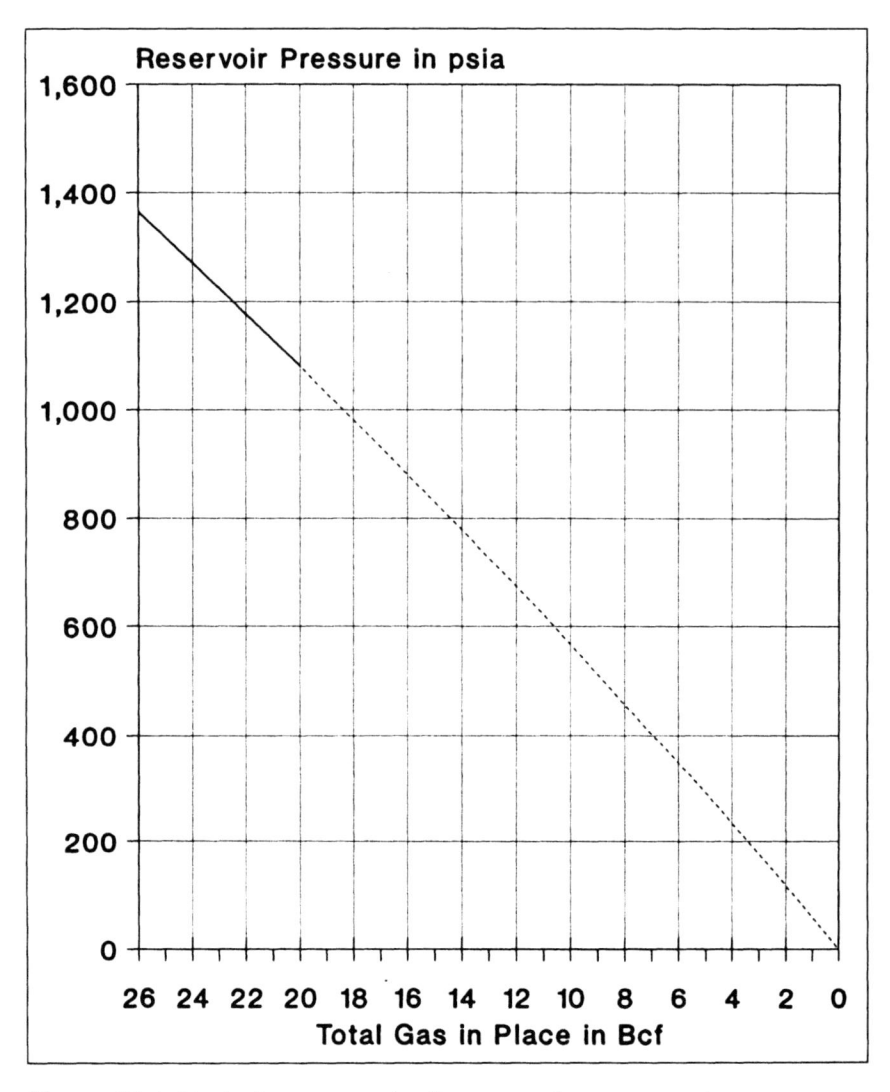

Figure 13-1. Typical pressure decline curve for a reservoir.

gas fields, some amount of extrapolation is necessary. A large amount of the error can be removed by plotting the cumulative withdrawals against P/Z rather than against pressure alone. The P/Z plot should approximate a straight line. One additional problem in obtaining data for this pressure decline curve is obtaining a stable pressure during the pressure test. When the field has been producing and is shut in, the pressure at each producing well will rise with time. The pressure will eventually stabilize, but it takes time. It is usually not economically feasible to shut in a producing field for a long period of time in order to obtain a stable pressure. Chapter 9 addresses this problem.

Four-point Flow Test. The second part of the deliverability forecasting procedure is to obtain the coefficients for the well flow equation for each well in the field. The well flow equation is:

$$Q = C \times (P_{SIS}^2 - P_{FS}^2)^n \qquad (13\text{-}1)$$

Q = flow rate of the well in Mcf per day
P_{SIS} = shut-in surface pressure in psia
P_{FS} = flowing surface pressure
C, n = constants that must be evaluated

The usual way to evaluate the constants C and n is to perform a four-point well test on each well. The field is shut in before this test to allow the pressures in the field to stabilize, and the shut-in surface pressure is recorded. The wellhead choke is then set for a given flow rate. This flow rate is held constant until the flowing surface pressure stabilizes. This can take many minutes or a few hours. Once the pressure and flow rate are stabilized, the data is recorded. The choke is then opened further to increase the flow rate. The second flow rate is held until the flowing surface pressure has again stabilized. The second set of data is recorded. This procedure is continued until four data points have been obtained. The data is then processed and plotted on a graph similar to Figure 13-2. The best straight line is drawn through the four points. This straight line is used to evaluate the constants C and n. C is the intercept when the value of Q is 1.0 and n is the slope of the line. A much better way to evaluate the constants is to take two points on the straight line and substitute each of them into the well flow equation. This gives two equations with two unknowns that can be solved for C and n.

Deliverability Forecast. Once the pressure decline curve for the reservoir has been established and the well flow equation coefficients have been evaluated for each well in the field, the future flow rates can be calculated. The reserves assigned to each well are divided into ten parts. This gives ten future production periods, with the production periods being of unequal time length. The length of time for the first production period will be relatively short, because the production rates are relatively high. The

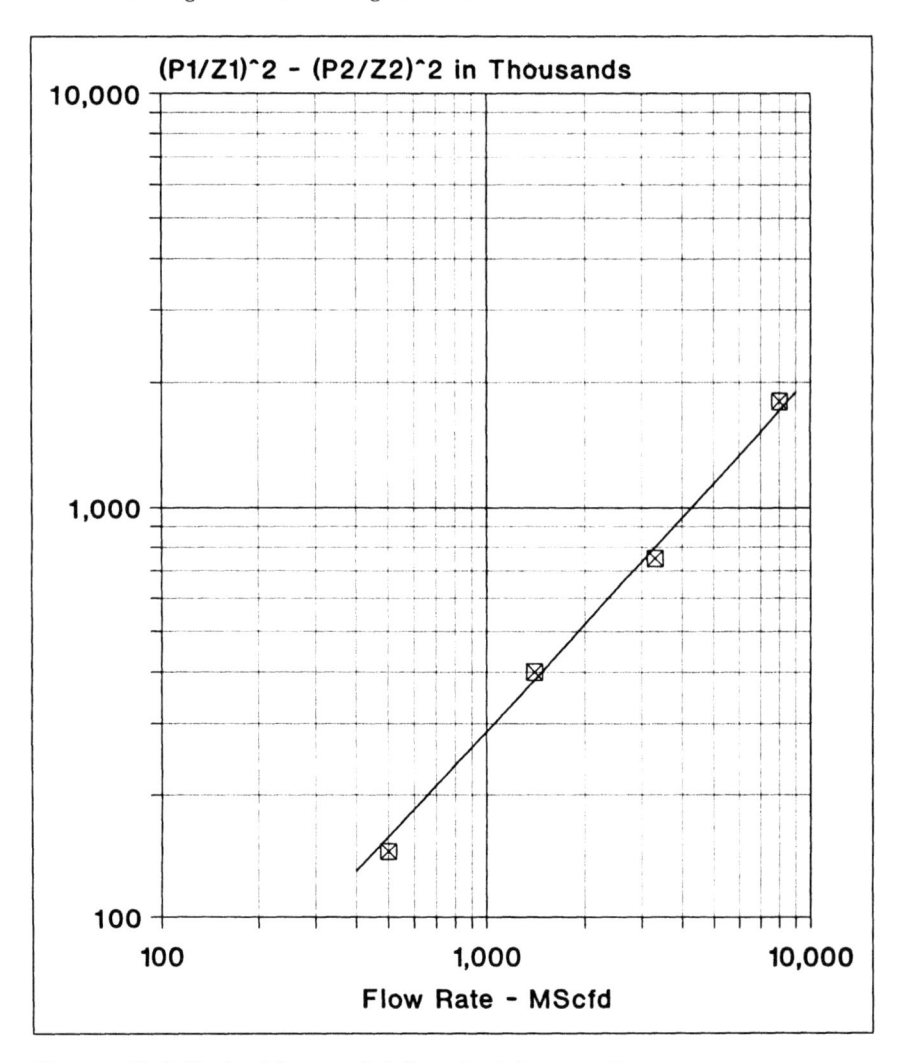

Figure 13-2. Typical four-point flow test for a well.

length of time for the tenth production period will be relatively long, as the reservoir is almost depleted, the reservoir pressure is low, and the production rate is low. The shut-in surface pressure at the start of each period is obtained from the pressure decline curve. A series of wellhead flowing pressures is assumed, and the flow rate for that well is calculated for each wellhead pressure. This is repeated for each production period. Table 13-1 gives an example of the output from this series of calculation. Figure 13-3 shows a typical deliverability decline curve.

Estimating Deliverability of Producing Wells **161**

Table 13-1
Example of Output from Petroleum Engineering Deliverability Study

Production Period	Volume During Period MMcf	SISP AT Start of Period psig	Wellhead Flowing Pressure psig	Peak Volume Mcfd
1	143	695	500	377
			450	455
			400	525
			350	586
			300	640
			250	686
			200	724
			150	754
			100	776
			50	791
2	143	640	500	260
			450	337
			400	407
			350	469
			300	523
			250	569
			200	607
			150	637
			100	659
			50	673
3	143	586	500	152
			450	229
			400	299
			350	361
			300	414
			250	460
			200	498
			150	528
			100	551
			50	565
4	143	431	500	53
			450	130
			400	200
			350	262
			300	316
			250	361
			200	399
			150	430
			100	452
			50	466

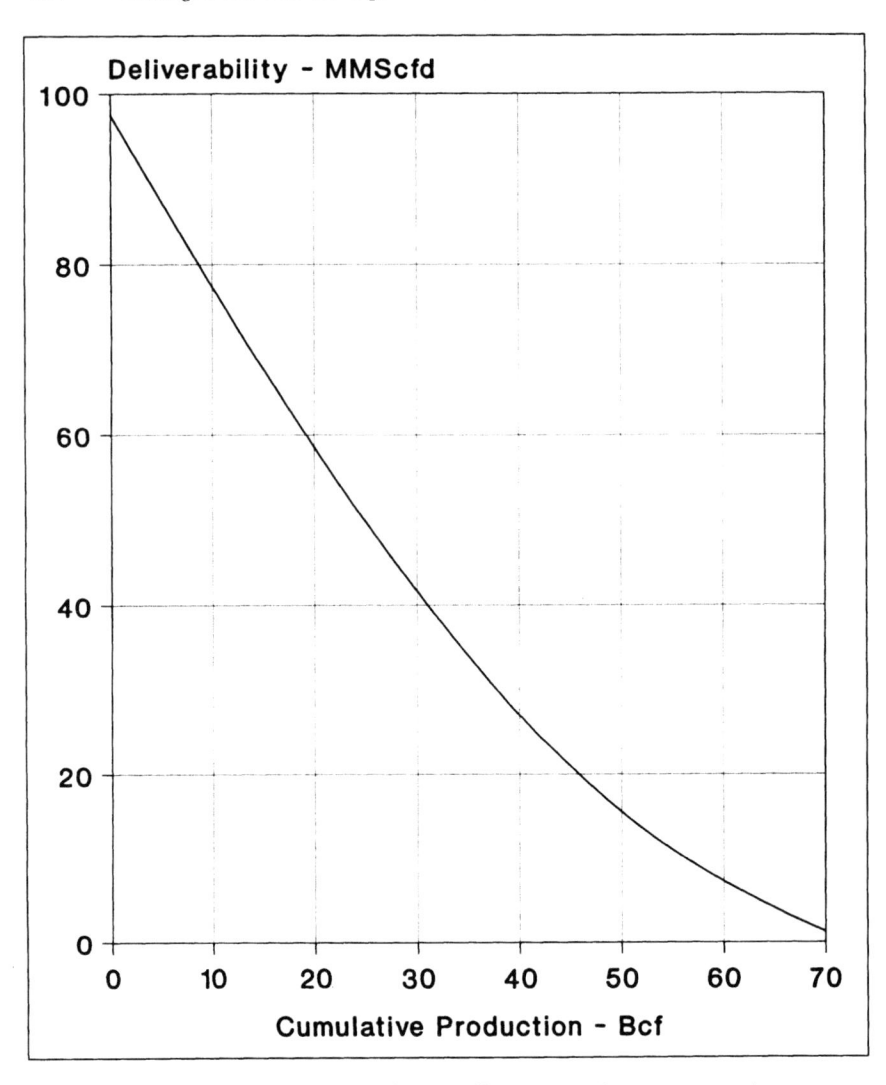

Figure 13-3. Typical deliverability decline curve for a reservoir.

Proposed Approximate Method

Theoretical Development. The procedure just described is a good procedure that gives good results, and is widely used in the industry. However, the data required is not always easily obtained. For this reason the question is sometimes asked as to whether there is a simpler method. What is needed is some method of relating the decline in deliverability to the production from the well or field. This raises the possibility of some type of decline factor that establishes this relationship. Flanigan [1] has suggested some possible relationships for this decline factor:

$$DF = -\frac{dD}{dQ} \qquad (13\text{-}2)$$

$$DF = -\frac{dD/D}{dQ} \qquad (13\text{-}3)$$

$$DF = -\frac{dD/D^a}{dQ} \qquad (13\text{-}4)$$

DF = decline factor
D = deliverability
Q = cumulative production from well or field
a = constant to be evaluated

An examination of these equations reveals that Equation 13-4 shows the most promise. Further, through the use of the constant "a" it actually embodies the features of the other two forms. If "a" is 0, Equation 13-4 becomes Equation 13-2. If "a" is 1.0, it becomes Equation 13-3. For any other value of "a", the Equation 13-4 retains its own form. When Equation 13-4 is integrated:

$$DF = \frac{1}{(1-a) \times \Delta Q} \times \left[D_1^{1-a} - D_2^{1-a}\right] \qquad (13\text{-}5)$$

This allows the decline factor to be determined from two data points.

By rearranging Equation 13-5 the deliverability at any point in the life of the reservoir can be obtained:

$$D = \left[D_1^{1-a} - (1-a) \times DF \times \Delta Q\right]^{\frac{1}{1-a}} \qquad (13\text{-}6)$$

Verification. In order to test this equation, the petroleum engineering deliverability study for a specific field was used for comparison. The particular field chosen was the Mather's Ranch field in the Texas panhandle. Two points in the early life of the field were used to calculate the decline factor DF. The points chosen were for cumulative production of 0 Bcf and 10 Bcf. The choosing of these points early in the life of the field gives a good test for the proposed method. A trial-and-error procedure was used to evaluate the

164 *Underground Gas Storage Facilities*

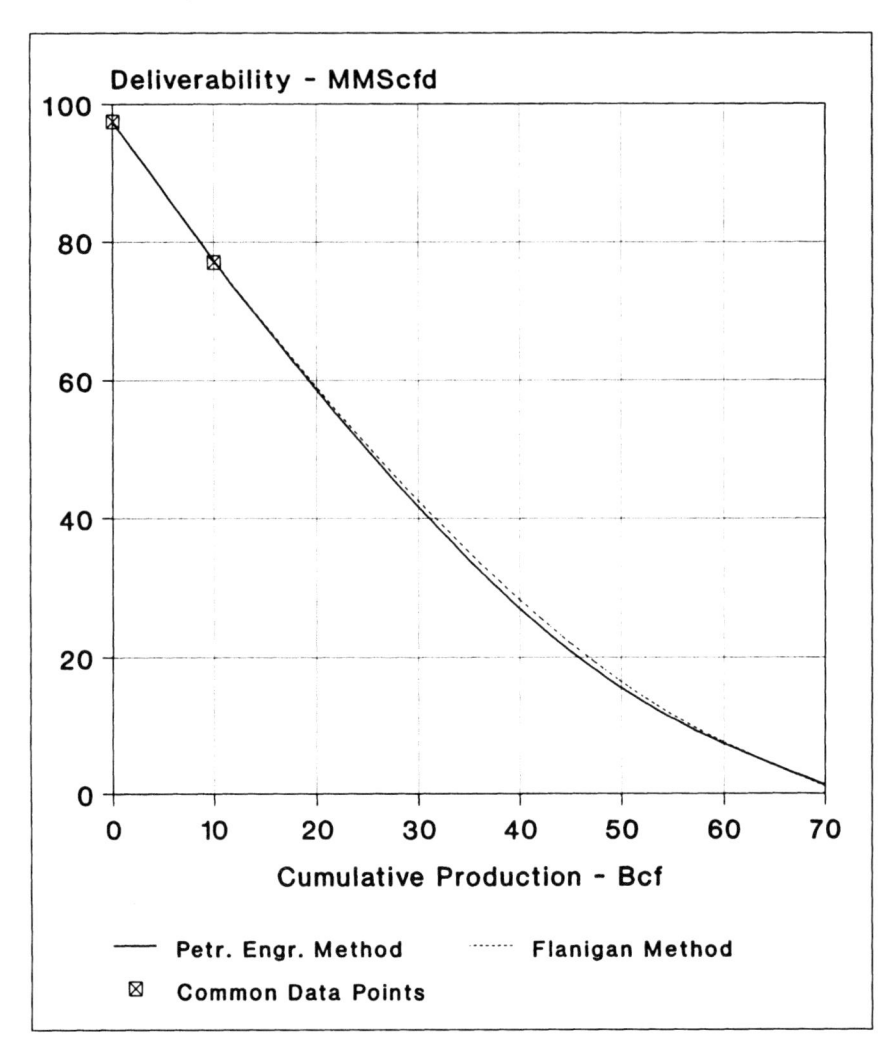

Figure 13-4. Decline curve for Mather's Ranch field with a flowing pressure of 600 psig.

constant "a", which gave the best fit and agreement with the petroleum engineering method. This resulted in a value of 0.386. The comparison with the petroleum engineering method is shown in Figure 13-4. The two data points at 0 and 10 Bcf cumulative production indicate the data points used to determine the decline factor. It can be seen that there is very good agreement throughout the entire range of the expected life of the property.

Estimating Deliverability of Producing Wells

This procedure was tried out on nine other wells and fields with results similar to Figure 13-4. The accuracy of the method appears to be acceptable. This method has also been used to forecast future deliverability of a company's entire field supply. The simplicity of the method is appealing in that the accuracy is acceptable and the data required is not onerous. Two data points are needed. One can be the deliverability test when the well is first put on stream. This makes a good data point because the reservoir is in equilibrium and stable as to pressure. If a second data point is available, that can be used. The FERC has a rule of thumb that 85 percent of the total reserves in a new reservoir are recoverable. This information can be used as a data point. Eighty-five percent of the total reserves would be the cumulative production and 0 would be the deliverability.

Sometimes it is desirable to know the deliverability in the future when the well is produced continuously at its maximum rate of flow. The incremental production with time would be:

$$dQ = D \times dt \tag{13-7}$$

where t is the time.
From Equation 13-4:

$$D^{-a} \times dD = -DF \times dQ \tag{13-8}$$

Combining Equations 13-7 and 13-8:

$$D^{-a} \times dD = -DF \times D \times dt \tag{13-9}$$

$$D^{-1-a} \times dD = -DF \times dt \tag{13-10}$$

Integrating:

$$\frac{1}{a} \times \left[D^{-a} - D_1^{-a} \right] = DF \times \Delta t \tag{13-11}$$

$$D = \left[D_1^{-a} + a \times DF \times \Delta t \right]^{-\frac{1}{a}} \tag{13-12}$$

This equation gives maximum deliverability with time. In order to determine cumulative production with time, Equations 13-7 and 13-12 may be combined:

$$dQ = D \times dt = \left[D_1^{-a} + a \times DF \times \Delta t\right]^{-\frac{1}{a}} \times dt \qquad (13\text{-}13)$$

Integrating:

$$\Delta Q = \frac{1}{(1-a) \times DF} \times \left[D_1^{1-a} - (D_1^{-a} + a \times DF \times \Delta t)^{\frac{a-1}{a}}\right] \qquad (13\text{-}14)$$

Equations 13-5, 13-8, 13-12, and 13-14 describe the decline of field reserves.

The question has been raised as to whether this method can be used to predict deliverability of storage wells. On storage wells, four-point flow tests have usually been run on each well. Reservoir pressure tests are usually run twice a year. Therefore the information is usually available to make a rigorous prediction of the performance of the storage well deliverability. The method proposed here could certainly be used as a check on the deliverability of storage wells.

References

1. Flanigan, Orin. *A Method of Predicting the Decline of Natural Gas Field Reserves.* Paper presented at the October, 1976 Annual Meeting of the Pipeline Simulation Interest Group.

CHAPTER 14

Automation

The original purpose of underground storage was to act as a supplemental supply when customer demands were beyond the limits of the normal sources of supply. On cold days the base storage was turned on and left on for a significant portion of the winter season. On very cold days, the peak storage facilities were turned on and left on until the very cold weather subsided. This might be several days later, and the peak storage was then turned off. Under these conditions the storage facilities could be manually operated with reasonable efficiency. The use of automation under these conditions was a luxury that not many companies made wide use of.

As manpower costs grew and gas control centers became more automated, it was a natural trend that automation of storage facilities would be increased. In addition, common usage developed a practice in storage facilities that was not planned for in the early facilities. As gas requirements go up and down with the weather and other conditions, it is necessary for the gas dispatcher to turn field supplies on and off and to make volume adjustments to gas supplies purchased from other pipelines. The turning on and off of a large number of wells to accommodate the weather patterns is a formidable chore. The gas dispatcher quickly discovered that it was considerably easier to leave the field supply on at a relatively constant volume and make the necessary adjustments by turning on or off a single source of supply, the underground storage. With the market conditions under which pipelines operate today, this practice is even more convenient and more prevalent. Some companies operate such that they inject into storage for part of a day and withdraw from storage another part of the same day. This practice has made automation and remote control of the storage facility highly desirable.

One usual practice for the operation of storage fields is to leave the injection-withdrawal wells connected to the gathering system at all times with the valves open. Some wells might have flow limiters on them during the withdrawal phase in order to better control conditions in the reservoir. In cases such as this, check valves are used to direct the gas to the proper path. With this method of operation, there is no need to operate

168 *Underground Gas Storage Facilities*

valves at the wellhead. All of the automation would then take place at the central point. The largest part of the automation effort is the automatic starting and stopping of the reciprocating compressors and operating the necessary valves to put these compressors on line and take them off line.

Reciprocating compressors have been successfully automated and remotely controlled since about 1961. Arkla and Mountain Fuel Supply were pioneers in this effort. Today many companies remotely start and stop reciprocating compressors and operate some of their facilities unattended. Many integral compressors are automated even though the station is manned. During this development of automation over a period of time some generally accepted philosophies have emerged.

During the early stages of automation of reciprocating compressors the prevailing attitude by operating companies was that the compressor manufacturers should design and manufacture the control equipment to start, stop, and monitor the engine-compressors while they were running. This attitude was engendered by the premise that manufacturers should know more about the peculiarities of their own equipment than anyone else. The compressor manufacturers today have their own electronic control divisions to continue this work, although there have been some non-compressor firms that have entered the field. The user specifies what features it wants in the start-up or shutdown sequence, and the control supplier designs these features into the standard package that has been developed.

The usual practice is to house the larger reciprocating compressors in a heated building. This is usually all that is needed if the engine is started manually. If the engine is started by remote control, it is good practice to have warm lubricating oil circulating to the engine. The reason for this disparity in recommendations is brought about by the procedure used when starting an automobile on a cold morning. In both cases of a manual start, the operator can listen to the engine and tell how long to crank and how long to warm the engine before loading it. In the case of a remote start, the operator relies entirely on the preset time intervals that are preprogrammed into the system. These intervals may not always be adequate for a cold engine, even in a heated building. The circulation of warm lubricating oil helps keep the engine warm and easier to start. The heater for the oil should have a flow switch that ensures that the oil is flowing before the heater is energized. Otherwise the heater can produce some carbonized residue in the oil system.

After start-up and warm-up of the engine, the automation system will operate the compressor and yard valves necessary to load the compressor and put it on line. While the engine is running there are a variety of operating functions that are monitored on a remotely operated compressor that are not monitored in a manned station. An example of this is a vibration monitor on the fin fans. In a manned station the fin fans are checked periodically. In a remotely operated station a severely unbalanced fan blade could do substantial damage to the equipment. The gap between the monitoring equipment installed in a manned station and in a remotely operated station is narrowing. The common use of some of this equipment has altered the price structure, and the operating and safety advantage that is gained by having the additional monitors has caused an increase in the sophistication of the equipment in manned stations.

The glycol dehydrator is another operating unit that sits in the cold for long periods of time and is then called to be put on the line in a short period of time. Although the starting of glycol dehydrators in cold weather does not present a large problem, some measures will make the effort easier. When glycol sits on the trays of the contactor in cold weather, the glycol tends to become a gel. One method to alleviate this is to have a small electric pump that circulates a small amount of glycol from the hot regenerator to the contactor to keep the vessel warm. A second potential problem area is the inlet valve to the dehydrator. If this valve is opened to put the dehydrator on line, the valve should be slow opening so as not to suddenly put a large surge of gas into the contactor. This surge can blow the glycol off the trays and down the pipeline and can even damage the trays.

APPENDIX A

Evaluation of Constants for Approximate Compressibility Factor Equation

Assume that a reservoir having a maximum capacity of 26.0 Bcf of total gas in place is being considered for gas storage or is being used for gas storage. It is anticipated that the cushion gas will approximate 12 Bcf, and working gas will be about 14 Bcf. Table A-1 lists the values for three points with different amounts of gas in the reservoir. The values of compressibility factors were taken from AGA Report #3. The equations that need constants evaluated are:

$$\text{GIP} = a + b \times \frac{P}{Z} \qquad \text{(A-1)}$$

$$Z = c + d \times P \qquad \text{(A-2)}$$

Table A-1
Selected Reservoir Characteristics

P psia	Z @ 110 °F	P/Z	GIP Bcf
1,365	0.8677	1,573.1	26.0
982	0.8979	1,093.7	18.0
568	0.9378	605.7	10.0

The values of GIP and P/Z for two of the points are substituted into the first equation:

$$26 = a + 1573.1 \times b \tag{A-3}$$

$$10 = a + 605.7 \times b \tag{A-4}$$

Subtracting Equation A-4 from Equation A-3:

$$16 = 967.4 \times b \tag{A-5}$$

$$b = 0.0165391$$

$$a = -0.0177793$$

Making similar substitutions into Equation A-2 and subtracting:

$$0.8677 = c + 1365 \times d \tag{A-6}$$

$$.9378 = c + 568 \times d \tag{A-7}$$

$$-0.0701 = 797 \times d \tag{A-8}$$

$$d = -0.000087955$$

$$c = 0.987759$$

These constants may be combined into the final equation:

$$\text{GIP} = -0.0177793 + \frac{0.0165392 \times P}{0.987759 - 0.000087955 \times P} \tag{A-9}$$

172   *Underground Gas Storage Facilities*

In order to check this equation, it can be used to calculate the GIP for the point in Table A-1 that was not used in the derivation of the constants:

$$\text{GIP} = -0.0177793 + \frac{0.0165392 \times 982}{0.987759 - 0.000087955 \times 982} \quad \text{(A-10)}$$

Calculated GIP = 18.001

The calculated GIP by the approximate method is 18.001 Bcf versus the actual value of 18.000 Bcf. This degree of accuracy is usually sufficient for engineering work on reservoirs. It should be kept in mind that the approximate equations should only be used within the conditions for which the constants were evaluated.

APPENDIX B

Development of Compressor Horsepower Equation

When compressing n mols of gas:

$$\Delta H = n \times \int_{T1}^{T2} C_p \times dT = n \times C_p \times (T_2 - T_1) \tag{B-1}$$

For a non-ideal gas [1]:

$$C_p - C_v = R \times Z \tag{B-2}$$

$$\frac{C_p}{C_v} = k \tag{B-3}$$

$$C_p = R \times Z + C_v = R \times Z + \frac{C_p}{k} \tag{B-4}$$

$$C_p = Z \times R \times \left(\frac{k}{k-1}\right) \tag{B-5}$$

Substituting Equation B-5 into Equation B-1:

$$\Delta H = n \times Z \times R \times \left(\frac{k}{k-1}\right) \times [T_2 - T_1] \tag{B-6}$$

The adiabatic expansion of a non-ideal gas may be expressed:

$$\frac{P_1 \times V_1^k}{Z_1} = \frac{P_2 \times V_2^k}{Z_2} \tag{B-7}$$

$$\frac{V_1}{V_2} = \left[\frac{P_2 \times Z_1}{P_1 \times Z_2}\right]^{1/k} \tag{B-8}$$

For any gas:

$$\frac{P_1 \times V_1}{T_1 \times Z_1} = \frac{P_2 \times V_2}{T_2 \times Z_2} \tag{B-9}$$

$$\frac{V_1}{V_2} = \frac{P_2 \times Z_1 \times T_1}{P_1 \times Z_2 \times T_2} = \left[\frac{P_2 \times Z_1}{P_1 \times Z_2}\right]^{1/k} \tag{B-10}$$

$$\frac{T_2}{T_1} = \frac{P_2 \times Z_1}{P_1 \times Z_2} \times \left[\frac{P_2 \times Z_1}{P_1 \times Z_2}\right]^{\frac{-1}{k}} \tag{B-11}$$

$$T_2 = T_1 \times \left[\frac{P_2 \times Z_1}{P_1 \times Z_2}\right]^{\frac{k-1}{k}} \tag{B-12}$$

Substituting Equation B-12 into Equation B-6:

$$\Delta H = n \times R \times Z_1 \times \left(\frac{k}{k-1}\right) \times \left[T_1 \times \left(\frac{P_2 \times Z_1}{P_1 \times Z_2}\right)^{\frac{k-1}{k}} - T_1\right] \tag{B-13}$$

$$\Delta H = n \times Z \times R \times T_1 \times \left(\frac{k}{k-1}\right) \times \left[\left(\frac{P_2 \times Z_1}{P_1 \times Z_2}\right)^{\frac{k-1}{k}} - 1\right] \tag{B-14}$$

$$P \times V = n \times Z \times R \times T \tag{B-15}$$

$$\Delta H = P_1 \times V_1 \times \left(\frac{k}{k-1}\right) \times \left[\left(\frac{P_2 \times Z_1}{P_1 \times Z_2}\right)^{\frac{k-1}{k}} - 1\right] \tag{B-16}$$

$$\frac{P_1 \times V_1}{Z_1 \times T_1} = \frac{Q \times P_s}{T_s} \tag{B-17}$$

$$P_1 \times V_1 = \frac{Q \times P_s \times T_1 \times Z_1}{T_s} \tag{B-18}$$

$$\Delta H = \frac{Z_1 \times Q \times P_s \times T_1}{T_s} \times \left(\frac{k}{k-1}\right) \times \left[\left(\frac{P_2 \times Z_1}{P_1 \times Z_2}\right)^{\frac{k-1}{k}} - 1\right] \quad \text{(B-19)}$$

Converting units:

$$HP = \frac{(1,000,000) \times (14.7) \times (144) \times Q \times T_1 \times Z_1}{(519.7) \times (24) \times (60) \times (33,000)} \times \left(\frac{k}{k-1}\right)$$

$$\times \left[\left(\frac{P_2 \times Z_1}{P_1 \times Z_2}\right)^{\frac{k-1}{k}} - 1\right] \quad \text{(B-20)}$$

$$HP = 0.085713786 \times Q \times T_1 \times Z_1 \times \left(\frac{k}{k-1}\right) \times \left[\left(\frac{P_2 \times Z_1}{P_1 \times Z_2}\right)^{\frac{k-1}{k}} - 1\right] \quad \text{(B-21)}$$

$$\frac{HP}{Q \times T_1 \times Z_1} = 0.085713786 \times \left(\frac{k}{k-1}\right) \times \left[\left(\frac{P_2 \times Z_1}{P_1 \times Z_2}\right)^{\frac{k-1}{k}} - 1\right] \quad \text{(B-22)}$$

Equations B-21 and B-22 are the theoretical equations for gas horsepower. They assume a pressure base of 14.7 psia and a temperature base of 60°F. The units for these equations are:

HP = horsepower
P = pressure in psia
T = temperature in °Rankine
Q = gas volume in MMScfd
Z = compressibility of gas
k = ratio of specific heats of gas

References

1. Dodge, B. F. *Chemical Engineering Thermodynamics*. New York: McGraw Publishing Company, 1944, p. 92.

Index

A

Absolute pressure, 32–3
Absorber, dehydrator, 134
AGA, 38, 124–5
American Gas Association, 38, 124–5
Automation, 167–9

B

Base load storage, 77–9
Boyle's law, 33–4

C

Capacity, working gas, 57–9, 72–5, 77–9
Charles' law, 34–5
Clearance pocket, compressor, 147
Colebrook-White correlation, 48
Company used gas, 19
Compressibility factor, 37, 38, 170–2
Compression ratio, 148
Compressor
　clearance pocket, 147
　compression ratio, 148
　control technology, 149
　cylinder unloaders, 147
　de-rating factor, 50
　diagnostic software, 151
　discharge temperature limitation, 148
　engines, 149
　fin fan coolers, 150–1
　lubrication options, 148–9
　maintenance, 151
　models, 151–6
　spare, 67
　separators, 149–50
　valve lifters, 147
　water cooled cylinders, 148
Compressors, 49–51
Conditioners, meter flow, 119, 121
Cooler, hydraulic fluid, 151,
Cooler sizing, 150
Coolers, fin fan, 150–1
Coolers, glycol, 137
Correlation, horsepower, 152–6
Cost of service, 84–9
Cushion gas, 44, 57
Cushion gas, effect on deliverability, 57–9
Customer growth rate, 13
Customer gas usage, 9, 16
Customers
　average annual number, 11
　average monthly number, 13–16
　large commercial, 10
　large industrial, 11
　medium industrial, 11
　residential, 10
　small industrial, 10
Cylinder unloaders, compressor, 147

D

Decline curve, deliverability, 160–2
Decline curve, pressure, 157–9
Decline factor, 163
Degree day, 16–18
Dehydrator
　absorber, 134
　economics, 137–9
　glycol cooler, 137
　glycol filter, 136

glycol pump, 135–6
problems, 139–40
regenerator, 136
still column, 136
stripping gas, 137, 140
turndown ratio, 139
types, 133–4
Dehydrators, glycol, 134
Dehydrators, solid bed, 133–4
Deliverability, 54–7, 157–66
Deliverability
decline factor, 163
effect of cushion gas volume, 57–9
forecast, 159–66
rating, 57
De-rating factor, compressors, 50
De-rating factor, wells, 62–3
Development, storage reservoir, 99–101
Dew point
effect of atmospheric temperature on, 141
hydrocarbon, 141
testers, 140–1
water, 140

E

Economics, dehydrator, 137–9
Economics of adding wells, 69–72, 88–9
Economics of storage, 63, 80–9
Energy
coal, 1
gas, 1
geothermal, 1
nuclear, 1
oil, 1
solar, 1
water power, 1
wind power, 1

F

Facilities, surface, 64–7
Fanning friction factor, 48
Field supply, 167
Filter, glycol, 136
Filter, pulsation, 122–4
Filter separators, 150
Forecast, deliverability, 159–66
Forecast, weather, 3
Four point well flow test, 45–7, 159
Frequency response, orifice meters, 128
Friction factor, 48

G

Gas, non-recoverable, 44, 54
Gas constant (R), 36–7
Gas holders, 3
Gas law, ideal, 36–8
Gas leakage, 90–108
Gas over-ride, 43–4, 91
Gas purchase contracts, 27–8
Gas Research Institute, 151
Gas supply, 26–7
Gathering system, 48–9
Gauge line error, 129–31
Gauge pressure, 32–3
Glycol
absorber, 134
contaminants, 140
cooler, 137
filter, 136
pumps, 135–6
purity, 136–7
regenerator, 136
still column, 136

H

Horsepower correlation, 152–6
Horsepower, theoretical, 152–3, 173–5 (B-1, B-3)
Hydrates, gas, 132
Hydrocarbon dew point, 141
HCPV, 101–8
Hydrocarbon pore volume, 101–8

I

Ideal gas law, 36–8
Injection wells, 44

K

Kelvin temperature scale, 33

L

Leakage, gas, 90–108
Leakage, types, 92
Leaks, large, 104–8
Load forecast, daily, 21–4
Lubrication options
    full lube, 148–9
    mini-lube, 148–9
    non-lube, 148–9

M

Maintenance, compressor, 151
Meter error, effect of pen position on, 119–21
Meter error, sources of, 119–21
Meter flow conditioners, 119, 121
Meter research, 121–2
Meter types, 118
Mid-range storage, 59–64
Models, compressor, 151–16
Moody diagram, 48

N

NIST, 121
Normal temperature patterns, 3
Number of customers
    adjustments to, 11–12
    average annual, 11
    average monthly, 13–16

O

Observation wells, 44, 48, 91
Optimization, 68–89
Orifice, square root error, 122
Orifice meter frequency response, 128
Orifice meter gauge line error, 129–31
Over-ride, gas, 43–4, 91

P

PCRC, 51, 122, 131, 151
PCRC design facility, 121–2
Peak load storage, 77–9
Pen position, effect on meter error, 119–21
Permeability, 40–1
Pipeline and Compressor Research Council, 51, 122, 131, 151
Porosity, 40
Pressure
    absolute, 32–3
    cycles, 92–108
    effect on volume, 33–4
    decline curve, 157–9
    gauge, 32–3
    measurement, 32–3, 109–17
    stabilization, 45, 94–9, 158–9, 109–17
Pressure-volume history, 92–108
Pulsating flow, 51, 121–31
Pulsating flow gauge line error, 129–31
Pulsation field tests, 122–6
Pulsation filter, 122–4
Pulsation Research Council, 121–2
Pump, glycol, 135–6

R

Rankine temperature scale, 33
Rate schedules, 19
Rating, deliverability, 57
Rating, surface facilities, 64–7
Rating of storages, 59–64
Regenerator, dehydrator, 136
Reservoirs, 40–44
Reservoir
    saddle, 44
    volumetric, 41–2, 94–5
    water drive, 41–2, 95–9

S

Saddle, geologic, 91
Separators, 149–50
SGA, 121
Slug catcher, 52
Southwest Research Institute, 121
Square root error, 122, 125–8
Square root error indicator, 130–31
SREI, 130–31
Still column, dehydrator, 136
Storage
   atmospheric pressure, 3
   base load, 59–64, 77–9
   characteristics, 28–9
   coal, 1–2
   economics, 63, 80–89
   mid-range, 59–64
   oil, 1–2
   peak load, 59–61, 77–9
   pressure vessels, 3
   rating, 59–64
   types, 59–64, 77–9
   wells, 44–8, 69–72
Storage reservoir development, 99–101
Stripping gas, 137, 140
Surface facilities, 64–7

T

Temperature,
   absolute, 33
   averaging, 4
   compressor discharge limitation, 148
   effect on dew point, 141
   effect on volume, 34–5
   Kelvin, 33
   measurement, 33
   patterns, 3
   profiles, 22–4
   Rankine, 33
   weighted average, 22–4
Temperature sensitive loads, 2, 10
Theoretical horsepower, 152–3, 173–5 (B-1, B-3)
Transmission lines, 53
Transported gas, 19
Turn down ratio, 139

U

USBM dew point tester, 140–1

V

Valve lifters, compressor, 147
Volumetric reservoir, 94–5

W

Water content charts, 142–3
Water content of natural gas, 133
Water content data, regression analysis of, 144–6
Water content specifications, 133
Water drive reservoir, 95–9
Weather Bureau data, 3, 4, 8
Weather forecast, 3
Well
   de-rating factor, 47, 62–3
   economics of adding, 69–72, 88–9
   injection-withdrawal, 44
   observation, 44, 48, 91
   withdrawal, 44
Wellhead, 61–4, 72–5, 77–9, 81–9
Working gas, 44
Working gas capacity, 57–9, 72–5, 77–9
Working gas volume, 72–9, 81–4
   effect on horsepower, 74–6
   effect on flowing wellhead pressure, 72–5, 77–9